Introduction to
General Relativity and Cosmology

Essential Textbooks in Physics

ISSN: 2059-7630

Published

Vol. 1   Newtonian Mechanics for Undergraduates
        by Vijay Tymms

Vol. 2   Introduction to General Relativity and Cosmology
        by Christian G. Böhmer

Essential Textbooks in Physics

Introduction to
General Relativity and Cosmology

Christian G. Böhmer
University College London, UK

World Scientific

NEW JERSEY · LONDON · SINGAPORE · BEIJING · SHANGHAI · HONG KONG · TAIPEI · CHENNAI · TOKYO

Published by

World Scientific Publishing Europe Ltd.

57 Shelton Street, Covent Garden, London WC2H 9HE

Head office: 5 Toh Tuck Link, Singapore 596224

USA office: 27 Warren Street, Suite 401-402, Hackensack, NJ 07601

Library of Congress Cataloging-in-Publication Data
Names: Böhmer, Christian G., author.
Title: Introduction to general relativity and cosmology / Christian G. Böhmer
    (University College London, UK).
Description: Covent Garden, London ; Hackensack, NJ : World Scientific,
    [2016] | Series: Essential textbooks in physics, ISSN 2059-7630 ; vol. 2
Identifiers: LCCN 2016023263| ISBN 9781786341174 (hc ; alk. paper) |
    ISBN 1786341174 (hc ; alk. paper) | ISBN 9781786341181 (pbk ; alk. paper) |
    ISBN 1786341182 (pbk ; alk. paper)
Subjects: LCSH: General relativity (Physics)--Textbooks. | Cosmology--Textbooks.
Classification: LCC QC173.6 .B64 2016 | DDC 530.11--dc23
LC record available at https://lccn.loc.gov/2016023263

British Library Cataloguing-in-Publication Data
A catalogue record for this book is available from the British Library.

Desk Editors: Suraj Kumar/Mary Simpson

Typeset by Stallion Press
Email: enquiries@stallionpress.com

Printed in Singapore

To Saffron, Isabelle, Lewis, Zachary, Martin and Wiera

Preface

Why another introduction to General Relativity and Cosmology? The primary motivation for this book was to present a concise yet detailed introduction to both General Relativity and Cosmology which first and foremost has the student reader in mind. When I was a student myself I very much disliked phrases like 'this is left as an easy exercise to the reader' or 'the reader may quickly verify that', and I also was not keen on exercises without any solutions or just some numerical answers. If, for instance, my calculation gave 7 and the answer was 6, how is it possible for a student to judge whether the entire calculation was fundamentally wrong or whether there was a small algebraic mistake, it might have been a typo in the book after all. With this in mind, the presentation of this book is fairly explicit, hopefully without being excessively concerned with minor details. Moreover, every chapter finishes with a short section called 'Further reading' which contains some references which would be a natural continuation for further studies.

When writing a textbook, there is always a temptation to add more material and to present detailed discussions and as many results as possible. On the other hand, students can easily feel intimidated by a tome covering the entire subject. From the students' point of view what is needed is a book with just the right amount of material. So, the aim was to keep the book within a page limit of approximately 260 pages and cover the standard material taught in two 10 week courses with 3 lectures per week. It was important to me to give detailed solutions to all 80 exercises which has taken up almost 60 pages of this book, so the actual main text is roughly 200 pages long which should be a manageable amount. The downside of this approach is that parts of this book might appear a little short on words, and that some interesting subjects had to be left out. Black holes are only mentioned in a few words and there was no room to cover the Kerr solution, other known exact solutions or the various different mass definitions used in General Relativity. In Cosmology it would have been wonderful to cover some cosmological perturbation theory and structure formation in detail, also Newtonian cosmology has some interesting features.

Chapter 1 covers differential geometry starting with standard vectors in Euclidean space. Chapter 2 begins with a brief discussion of some Physics topics, and then motivates the structure of the Einstein field equations of General Relativity. The weak field approximation and gravitational waves are discussed. The chapter finishes with deriving the Einstein field equations by using the variational approach which is particularly elegant. The 'Further reading' section covers some material related to modifications and extension of Einstein's theory. In Chap. 3, the Schwarzschild solutions are discussed. This includes the classical Solar System tests of General Relativity, all of which are in excellent agreement with the theoretical predictions. An introduction to Cosmology is given in Chap. 4. This part includes inflation and should take students to the beginnings of what one now calls Modern Cosmology. As already stated, the final Chap. 5 contains the fully worked out solutions to all exercises.

I am keen to get input and feedback from readers. Therefore, there is a webpage http://book.christianboehmer.co.uk/ where I will post readers' suggestions, keep a list of typos and corrections, etc. I will also make some Mathematica files available which can be used to compute the various tensors needed in General Relativity and Cosmology. I will also keep a list of additional topics suggested by readers.

It is difficult to mention each and every one who has influenced this book. Therefore, I would like to thank all my collaborators, and colleagues at UCL for the many useful discussions we have had over the years. I would like to thank Laurent Chaminade from World Scientific who motivated me to start this project. I am also very grateful to Sebastian Bahamonde, Nyein Chan, Saffron Glenister, Atifah Mussa, Nicola Tamanini and Matthew Wright for reading through the manuscript and pointing out various typos, inaccuracies and making suggestions for improvements.

London & Potton, February 2016

C. G. Boehmer

About the Author

Dr. Christian Böhmer is a Reader in Mathematics at University College London where he teaches various undergraduate and graduate courses. He is also involved in teaching PhD students at the London Taught Course Centre. Dr. Böhmer studied physics at the University of Potsdam, the Technical University of Berlin and University College Dublin. After graduation, he completed his PhD studies at the Vienna University of Technology. His research interests are in general relativity and its modifications, cosmology, and aspects of continuum mechanics. He has over 60 peer-reviewed articles and has presented his work in many seminars, and at national and international conferences.

Contents

1
Differential Geometry

The theory of General Relativity is a theory of gravitation based on the geometric properties of spacetime. Its formulation requires the use of differential geometry. One of the great difficulties when working with geometric objects on arbitrary spaces is notation. As far as Cartesian tensors are concerned, the issue is much easier. However, in order to prepare for the later parts on Riemannian and Lorentzian geometry, we will introduce most of the abstract notation of differential geometry in the familiar Euclidean setting.

1.1. The Concept of a Vector

Let us start with Euclidean space denoted by $\mathbb{E}^3$. A vector v is a quantity in $\mathbb{E}^3$ with specified direction and magnitude. The magnitude of this vector is denoted either by $|v|$ or v. Graphically we can view a vector as an arrow, with its length representing the magnitude. Various physical quantities are best represented by vectors, examples are forces, velocities, moments, displacements and many others. We define the zero vector O as a vector with zero magnitude and arbitrary direction. For any vector v, we can define a unit vector pointing in the same direction as v but with magnitude 1, simply by $\hat{v} = v/|v|$. Two vectors are called equal if they have the same direction and the same magnitude.

The concept of a vector as such does not require the introduction of coordinates. This is important conceptually, the outcome of an experiment should not depend on our choice or coordinates. In most applications, however, a choice of good coordinates which are adapted to the physical system can considerably simplify subsequent equations. Try for instance solving the two simple equations $\ddot{x} = 0$, $\ddot{y} = 0$ using polar coordinates, it becomes difficult.

However, choosing coordinates and units can also lead to substantial problems. In particular, when two groups working on the same project assume they use the same coordinates. It was exactly this assumption that made NASA's Mars Climate Orbiter mission a failure. Subcontractor Lockheed Martin designed thruster software that used Imperial units, while NASA uses metric units in their software. The spacecraft approached Mars at a much lower altitude than expected and it is likely that atmospheric stresses destroyed it (NASA, 1999). As trivial as the matter seems to appear, units, coordinates and their transformation properties are a crucial ingredient of Engineering, Mathematics and Physics. It also explains why theoretical physicists are keen to set every possible constant to one.

1.1.1.   *Vector operations*

Having introduced the concept of a vector, we must next define admissible algebraic operations on vectors. Let u and v be two vectors, then we define their sum $u + v$ to be the vector which completes the triangle when the tail of v is placed at the tip of u. When u and v are interchanged, one arrives at the same vector (parallelogram rule) and thus $u + v = v + u$. We can also multiply the vector v by an arbitrary real number $r \in \mathbb{R}$ to get the vector rv. This vector has the same direction as v if $r > 0$ but has opposite direction if $r < 0$ and it has length $|r||v|$. When $r = 0$, we get the zero vector. Addition and the notion of the zero vector allow us to define the inverse (under addition) of a vector by saying that $\tilde{v}$ is the inverse of v if $\tilde{v} + v = \mathbf{0}$. We will denote this simply by $\tilde{v} = -v$. We can now state what is meant by the operation $u - v$, namely this vector obtained by placing the tip of v at the tip of u.

Two vectors uniquely define a plane. Let us move the vectors such that their tails join at the same point. Now, in this plane we can define the angle θ, say, between these two vectors. This angle can also be used to measure the area of the parallelogram spanned by the vectors. The angle between any two vectors should be independent of the vectors' lengths and should only depend on their orientation. Let us define the scalar product between two vectors $\boldsymbol{u}$ and $\boldsymbol{v}$ by $\hat{\boldsymbol{u}} \cdot \hat{\boldsymbol{v}} = \cos\theta$. Using $\hat{\boldsymbol{v}} = \boldsymbol{v}/|\boldsymbol{v}|$, this can be rewritten into the standard form of the scalar product

$$\boldsymbol{u} \cdot \boldsymbol{v} = |\boldsymbol{u}||\boldsymbol{v}| \cos\theta \,, \tag{1.1}$$

where $\theta \in [0, \pi]$. Two vectors are said to be perpendicular or orthogonal if $\boldsymbol{u} \cdot \boldsymbol{v} = 0$. Note that $\boldsymbol{v} \cdot \boldsymbol{v} = |\boldsymbol{v}|^2$.

Having specified a plane spanned by two vectors, we can alternatively define the same plane by only one (different) vector. Let us take a vector $\boldsymbol{w}$ which is orthogonal to both $\boldsymbol{u}$ and $\boldsymbol{v}$. Then this vector can be used to define the plane spanned by $\boldsymbol{u}$ and $\boldsymbol{v}$. If we combine this with the idea of measuring the area of the parallelogram spanned by the vectors, we arrive at the definition of the vector product

$$\boldsymbol{u} \times \boldsymbol{v} = |\boldsymbol{u}||\boldsymbol{v}| \sin\theta \, \boldsymbol{e} \,, \tag{1.2}$$

where $\boldsymbol{e}$ is a unit vector $(|\boldsymbol{e}| = 1)$ orthogonal to both $\boldsymbol{u}$ and $\boldsymbol{v}$, and normal to the plane spanned by them. Moreover, the orientation of $\boldsymbol{e}$ is given by the right-hand rule. This means $\boldsymbol{u}$, $\boldsymbol{v}$ and $\boldsymbol{e}$ form a right-handed set. Should we use a left-hand rule, then this would introduce an additional minus sign in the definition. The length $|\boldsymbol{u} \times \boldsymbol{v}|$ gives the area of the parallelogram. Note that $\boldsymbol{u} \times \boldsymbol{v} = -\boldsymbol{v} \times \boldsymbol{u}$.

1.1.2. *Projections and basis vectors*

The scalar product has a second geometrical interpretation based on the idea of projections. Let us project $\boldsymbol{u}$ onto $\boldsymbol{v}$ and denote the projected vector by $\boldsymbol{u}_v$. The length of the projection is given by

$|\boldsymbol{u}_v| = |\boldsymbol{u}| \cos \theta$ which can be written using the scalar product as $|\boldsymbol{u}_v| = \boldsymbol{u} \cdot \hat{\boldsymbol{v}}$. Since this projection points in the direction of $\boldsymbol{v}$, the projected vector is given by $\boldsymbol{u}_v = (\boldsymbol{u} \cdot \hat{\boldsymbol{v}})\hat{\boldsymbol{v}}$ which we write in the form

$$\boldsymbol{u}_v = \frac{\boldsymbol{u} \cdot \boldsymbol{v}}{\boldsymbol{v} \cdot \boldsymbol{v}} \boldsymbol{v}. \tag{1.3}$$

Therefore, the vector $\boldsymbol{u}$ can be decomposed into two components with respect to $\boldsymbol{v}$, $\boldsymbol{u} = \boldsymbol{u}_v + \boldsymbol{u}_v^\perp$. One part, $\boldsymbol{u}_v$, which is the projection of $\boldsymbol{u}$ onto $\boldsymbol{v}$, and another part orthogonal to $\boldsymbol{u}_v$ which is given by

$$\boldsymbol{u}_v^\perp = \boldsymbol{u} - \frac{\boldsymbol{u} \cdot \boldsymbol{v}}{\boldsymbol{v} \cdot \boldsymbol{v}} \boldsymbol{v}. \tag{1.4}$$

In order to see that $\boldsymbol{u}_v$ and $\boldsymbol{u}_v^\perp$ are orthogonal, it suffices to compute

$$\begin{aligned}
\boldsymbol{u}_v \cdot \boldsymbol{u}_v^\perp &= [(\boldsymbol{u} \cdot \hat{\boldsymbol{v}})\hat{\boldsymbol{v}}] \cdot [\boldsymbol{u} - (\boldsymbol{u} \cdot \hat{\boldsymbol{v}})\hat{\boldsymbol{v}}] \\
&= (\boldsymbol{u} \cdot \hat{\boldsymbol{v}})(\boldsymbol{u} \cdot \hat{\boldsymbol{v}}) - (\boldsymbol{u} \cdot \hat{\boldsymbol{v}})^2 (\hat{\boldsymbol{v}} \cdot \hat{\boldsymbol{v}}) = 0. \tag{1.5}
\end{aligned}$$

Let us assume that $\boldsymbol{i}$ is a given unit vector. Then, we can uniquely decompose the vector $\boldsymbol{v}$ into $\boldsymbol{v}_i$ and $\boldsymbol{v}_i^\perp$. Next, we consider three mutually orthogonal unit vectors $\boldsymbol{i}, \boldsymbol{j}, \boldsymbol{k}$ which satisfy $\boldsymbol{i} \times \boldsymbol{j} = \boldsymbol{k}$, $\boldsymbol{j} \times \boldsymbol{k} = \boldsymbol{i}$ and $\boldsymbol{k} \times \boldsymbol{i} = \boldsymbol{j}$, we refer to such a set as a basis. Then we can decompose $\boldsymbol{v}$ uniquely with respect to those three unit vectors by repeated projection along the three given vectors

$$\boldsymbol{v} = (\boldsymbol{v} \cdot \boldsymbol{i})\boldsymbol{i} + (\boldsymbol{v} \cdot \boldsymbol{j})\boldsymbol{j} + (\boldsymbol{v} \cdot \boldsymbol{k})\boldsymbol{k}. \tag{1.6}$$

We refer to the numbers $(\boldsymbol{v} \cdot \boldsymbol{i})$, $(\boldsymbol{v} \cdot \boldsymbol{j})$ and $(\boldsymbol{v} \cdot \boldsymbol{k})$ as the components of the vector $\boldsymbol{v}$ in the basis $\{\boldsymbol{i}, \boldsymbol{j}, \boldsymbol{k}\}$. Note that it is important to distinguish the representation of a vector (three real numbers) in a given basis from its abstract meaning (a directed line segment).

Given a basis $\{\boldsymbol{i}, \boldsymbol{j}, \boldsymbol{k}\}$, we can now compute all possible combinations of scalar products (1.1) which are

$$\begin{array}{lll}
\boldsymbol{i} \cdot \boldsymbol{i} = 1, & \boldsymbol{i} \cdot \boldsymbol{j} = 0, & \boldsymbol{i} \cdot \boldsymbol{k} = 0, \\
\boldsymbol{j} \cdot \boldsymbol{i} = 0, & \boldsymbol{j} \cdot \boldsymbol{j} = 1, & \boldsymbol{j} \cdot \boldsymbol{k} = 0, \\
\boldsymbol{k} \cdot \boldsymbol{i} = 0, & \boldsymbol{k} \cdot \boldsymbol{j} = 0, & \boldsymbol{k} \cdot \boldsymbol{k} = 1.
\end{array} \tag{1.7}$$

Clearly, whenever two different basis vectors come together, their scalar product returns zero as any two such vectors are perpendicular. On the other hand, when the scalar product between a basis vector and itself is computed, the angle is zero, giving one. Recall that basis vectors are normalised.

An important observation is that the right-hand sides of (1.7) look like the 3×3 identity matrix, a point which helps to justify the more compact index notation which we will introduce shortly.

When the possible vector products (1.2) are computed we find

$$\begin{aligned}
\boldsymbol{i} \times \boldsymbol{i} = 0, \qquad & \boldsymbol{i} \times \boldsymbol{j} = \boldsymbol{k}, \qquad & \boldsymbol{i} \times \boldsymbol{k} = -\boldsymbol{j}, \\
\boldsymbol{j} \times \boldsymbol{i} = -\boldsymbol{k}, \qquad & \boldsymbol{j} \times \boldsymbol{j} = 0, \qquad & \boldsymbol{j} \times \boldsymbol{k} = \boldsymbol{i}, \\
\boldsymbol{k} \times \boldsymbol{i} = \boldsymbol{j}, \qquad & \boldsymbol{k} \times \boldsymbol{j} = -\boldsymbol{i}, \qquad & \boldsymbol{k} \times \boldsymbol{k} = 0.
\end{aligned} \qquad (1.8)$$

Here we should take notice that the right-hand sides look like a skew-symmetric (anti-symmetric) matrix. The sign of the right-hand side of the equation can be determined by noting that it is $+$ when the order of the three vectors forms an even permutation of $\boldsymbol{i}, \boldsymbol{j}, \boldsymbol{k}$ and it is $-$ when the permutation is odd.

1.1.3. *Towards tangent space and all that*

Before introducing the very useful index notation, it seems appropriate to briefly present a more abstract view of vectors, we will make this more formal later. This viewpoint will also be useful when we discuss Riemannian geometry or simply curved spaces. Let us consider the 2-sphere, or plainly the surface of the Earth. Pick two points and connect them with an arrow (straight line with orientation) from A to B, say. It is now tempting to refer to this as a vector, however, it is not. Firstly, if we carry out the aforementioned procedure honestly, then this straight line has to penetrate the surface of the Earth and go through parts of its interior. On the other hand,

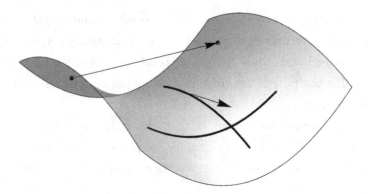

Fig. 1.1   Shown is an arbitrary surface. The arrow which connects the two points does not lie within this surface and hence cannot be regarded as a vector. Also shown are two curves on this surface and a tangent vector to one of the curves. Tangent vectors are the building blocks of tangent space.

if we insist on drawing the line on the surface, it would not be a straight line any more, it would be part of a great circle, provided we choose the shortest possible line connecting A and B. Thus, we seem to struggle with applying the concept of a vector to the situation of the sphere, or any other curved surface, see also Fig. 1.1. Yet, everyday life seems to suggest that our approximation of the surface of the Earth as flat Euclidean space is working rather well locally. The reason for this is that at every point of the 2-sphere we can define the plane tangent to its surface. This plane is Euclidean and locally, this means in a small neighbourhood around this point, every space looks Euclidean. We will not consider spaces with corners or sharp edges.

It is a very convenient coincidence that in Euclidean three space $\mathbb{E}^3$, the space and its tangent space at every point are isomorphic, this means their structures are the same. In simple words, a tangent vector to a plane is itself part of this plane. The above basis vectors $\{\boldsymbol{i}, \boldsymbol{j}, \boldsymbol{k}\}$ should, strictly speaking, be viewed as the basis elements of the tangent space. Then there exists a dual tangent space with a dual basis. The dual space consists of mappings which take elements of the tangent space and maps them to the real numbers. This is nothing but the scalar product (1.1), which is then interpreted as an inner product. In Euclidean space $\mathbb{E}^3$, the space itself, its tangent space and the dual tangent space are all isomorphic, in general this is not the case. However, there will always be a mapping which allows us to map elements of the tangent space to its dual space. This mapping will turn out to be the metric. It is the metric which is of great importance for applications of physics. It is the object which determines how we measure distances.

While we do not require this abstract setting yet, it helps to keep this in mind as it will make the subsequent introduction of the index notation clearer.

1.1.4. *Index notation*

We have seen that every vector v in a Euclidean space can be decomposed with respect to its basis vectors, see Eq. (1.6). This procedure can easily be extended to n dimensions in which case we would have n basis vectors. Let us start with an arbitrary set of basis vectors $\{e_1, e_2, \ldots, e_n\}$ which we will denote by e_i, $i = 1, \ldots, n$. As mentioned in the previous subsection, this is a basis for the tangent space. Let us denote the elements of the dual basis by e^j, $j = 1, \ldots, n$, the e^j are the basis for the dual tangent space. These two basis are dual in the sense that

$$
e_i \cdot e^j = \delta_i^j = \begin{cases} 1 & \text{if } i = j, \\ 0 & \text{if } i \neq j, \end{cases} \tag{1.9}
$$

where the meaning of the scalar product is as before and we introduced the Kronecker delta δ_i^j. One can think of δ_i^j as the components of the identity matrix, see the remark after (1.7).

We write a vector $\boldsymbol{v}$ in the basis $\boldsymbol{e}_i$ in the form

$$\boldsymbol{v} = v^1\boldsymbol{e}_1 + v^2\boldsymbol{e}_2 + \cdots + v^n\boldsymbol{e}_n \tag{1.10}$$

$$= \sum_{i=1}^{n} v^i\boldsymbol{e}_i. \tag{1.11}$$

We refer to the v^i as the contravariant components of the vector $\boldsymbol{v}$ in the basis $\boldsymbol{e}_i$. Summations over repeated indices will occur at many places in most of the subsequent equations. We thus introduce the Einstein summation convention whereby one suppresses the summation symbol and sums over twice repeated (one upper and one lower) indices. We simply write

$$\boldsymbol{v} = v^i\boldsymbol{e}_i, \tag{1.12}$$

to mean the same as Eq. (1.11). Since the notion of a vector is independent of our choice of basis, we could equivalently write

$$\boldsymbol{v} = v_i\boldsymbol{e}^i, \tag{1.13}$$

where we call v_i the covariant components of the vector. Note that in the Euclidean $\mathbb{E}^3$ the numbers v^i and v_i are identical. Since $\mathbb{E}^3$, its tangent space and the dual tangent space are all isomorphic to $\mathbb{E}^3$, one does not have to differentiate between an upper and a lower index.

The previously defined scalar product between two vectors $\boldsymbol{u}$ and $\boldsymbol{v}$ can now be written as follows:

$$\boldsymbol{u} \cdot \boldsymbol{v} = (u^i\boldsymbol{e}_i) \cdot (v_j\boldsymbol{e}^j) = u^i v_j(\boldsymbol{e}_i \cdot \boldsymbol{e}^j)$$
$$= u^i v_j \delta_i^j = u^i v_i = u_i v^i. \tag{1.14}$$

As expected, the scalar product is independent of the choice of our basis. We also note that it easily extends to n dimensions.

At this point we will introduce the Levi-Civita symbol which is defined by

$$\varepsilon_{ijk} = \begin{cases} +1 & \text{if } ijk = 123, 312, 231, \\ -1 & \text{if } ijk = 213, 132, 321, \\ 0 & \text{otherwise.} \end{cases} \tag{1.15}$$

This can be used to define the vector product using the index notation

$$\boldsymbol{u} \times \boldsymbol{v} = (u^i \boldsymbol{e}_i) \times (v^j \boldsymbol{e}_j) = \varepsilon_{ijk} u^i v^j \boldsymbol{e}^k. \tag{1.16}$$

Alternatively, we could state $\boldsymbol{e}_i \times \boldsymbol{e}_j = \varepsilon_{ijk} \boldsymbol{e}^k$. A crucial point to note here is the position of the vectors. The object $\boldsymbol{e}_i \times \boldsymbol{e}_j$ should really be viewed as an area element and not a vector as such. In three dimensions, areas and vectors are dual objects in the sense that a single vector can uniquely characterise an area, see the discussion after Eq. (1.16). The direction of the vector determines the orientation of the area and the length of the vector determines its size.

1.2. Manifolds and Tensors

1.2.1. *Tangent space and vector fields*

We start with an n-dimensional vector space V with coordinates $\{X^1, \ldots, X^n\}$ such that V is an open subset of $\mathbb{R}^n$. If we consider the surface of the sphere $\mathbb{S}^2$, defined by the condition $(X^1)^2 + (X^2)^2 + (X^3)^2 = 1$ then, as discussed earlier, we cannot easily define vectors which are part of this space. In this example we view $\mathbb{S}^2$ embedded in $\mathbb{R}^3$, the latter being a vector space.

In what follows we will need the notion of a smooth function f. This is a function $f : V \mapsto \mathbb{R}$ whose partial derivatives of any order exist. This means we can differentiate f with respect to any of its variables as many times as needed. The space of all such functions is denoted by $C^\infty(V)$.

Definition 1.1 (Curve). A curve C is a smooth mapping from an interval $C : I \mapsto V$ where $I \subset \mathbb{R}$. We generally write the curve C as $X^i = X^i(\tau)$ where τ is the parameter of the curve.

When we think about standard calculus on the real line, we can find the derivative of a function (provided it exists) and use the value of the derivative at a given point to construct the tangent at this point. We are now extending this idea to higher dimensions.

Definition 1.2 (Tangent vector of a curve). The tangent vector is given by $T = T^i e_i$ where the components T^i of this vector are given by $T^i = \frac{dX^i}{d\tau}$.

This is quite an intuitive definition in agreement with Newtonian mechanics. We can think of $X^i(\tau)$ as the trajectory of a particle and $T^i(\tau)$ would then correspond to the particle's velocity. We are tempted to write $\dot{X}^i(\tau)$. In this case one naturally identifies τ with Newtonian time. It is clear that the components T^i depend on the choice of coordinates in V, however, the tangent vector to a curve should be independent of that choice.

We could have chosen a different parameter, λ say, for our curve C so that $X^i(\lambda(\tau))$. Then the tangent vector to the curve is

$$T^i = \frac{dX^i}{d\tau} = \frac{dX^i}{d\lambda}\frac{d\lambda}{d\tau}. \tag{1.17}$$

It is possible to always parametrise a curve C such that the tangent vector to this curve has unit length.

Definition 1.3 (Affine parametrisation). The parameter λ of a curve C is called affine if the tangent vector to this curve has unit length.

Example 1.1 (Affine parametrisation). Consider the curve $(x, y) = (\sqrt{1 - t^2}, t)$. The tangent vector to this curve is given by $\boldsymbol{T} = (-t/\sqrt{1 - t^2}, 1)$ and it has length $1/\sqrt{1 - t^2}$. Let us define a new parameter s by $t = \sin(s)$, in this new parameter our curve is given by $(x, y) = (\cos(s), \sin(s))$. In this parametrisation the tangent vector is $\boldsymbol{T} = (-\sin(s), \cos(s))$ which has unit length. s is nothing but the arc length of the curve.

Definition 1.4 (Tangent space T_pV). Let p be a point in V and consider all possible curves C which contain p. We define the tangent space T_pV to be the set of all tangent vectors of curves at p.

In Euclidean space, we are used to add and subtract vectors at different points, as these vectors were all constant. In curved spaces, vectors of the same tangent space at some point p can be combined, however, vectors at different points are independent objects.

Our next question is the following. Our space V came equipped with some coordinates X^i which suggests that there should exist some natural basis vectors $\boldsymbol{e}_i$ for the tangent of a curve. In particular, this choice should be such that the vector $\boldsymbol{T}$ is invariant under coordinate changes. Let f be a smooth function on V and consider the object $F = f(X^i(\tau))$ which is a mapping $F : \mathbb{R} \mapsto \mathbb{R}$, or simply a function on the real line.

Let us use the chain rule to compute the derivative of F with respect to τ, we find

$$\frac{dF}{d\tau} = \frac{\partial f}{\partial X^i} \frac{dX^i}{d\tau} = T^i \frac{\partial}{\partial X^i} f. \tag{1.18}$$

Since this is true for all smooth f, we are led to identify the basis vectors $\boldsymbol{e}_i$ with the partial derivatives $\partial/\partial X^i$. Note that this object corresponds to the directional derivative of the function f along the direction T^i. In modern differential geometry one speaks of the vector field $\boldsymbol{T}$ which is a mapping from the smooth functions into the tangent space. In vector calculus one would write $\boldsymbol{T} \cdot \nabla f$. To check if this vector $\boldsymbol{T}$ has indeed the desired properties we first define what is meant by a coordinate transformation.

Definition 1.5 (Coordinate transformations). Consider the vector space V with coordinates $\{X^1, \ldots, X^n\}$. Consider a set $\{Y^1, \ldots, Y^n\}$ such that the Y^i are smooth functions of the old coordinates $Y^1(X^1, \ldots, X^n)$, $Y^2(X^1, \ldots, X^n)$, etc. For $\{Y^1, \ldots, Y^n\}$ to be a coordinate system of V, it is required that the matrix of first derivatives

$$J^i{}_j = \frac{\partial Y^i}{\partial X^j}, \tag{1.19}$$

the Jacobian, is invertible.

Let us now return to the tangent vector $\boldsymbol{T}$. The concept of this vector should be independent of the choice of coordinates. So, let us consider two coordinate systems $\{X^1, \ldots, X^n\}$ and $\{Y^1, \ldots, Y^n\}$. Using the chain rule we have

$$\frac{\partial}{\partial X^i} f = \frac{\partial Y^j}{\partial X^i} \frac{\partial}{\partial Y^j} f. \tag{1.20}$$

Therefore, the partial derivatives of f change under coordinate transformations. For the vector $\boldsymbol{T}$ to be invariant under coordinate transformations, it is required that the components T^i transform inversely to the partial derivatives, this means we require

$$T_X^i = \frac{\partial X^i}{\partial Y^j} T_Y^j, \tag{1.21}$$

where the subscripts X, Y indicate that the components T^i are those corresponding to the coordinates X^i and Y^i, respectively. This is a somewhat involved notation which is only used in the following calculation to emphasise the underlying mathematics. In this case we have

$$\boldsymbol{T} = T^i_{(X)} \boldsymbol{e}_{i(X)} = T^i_{(X)} \frac{\partial}{\partial X^i} = \frac{\partial X^i}{\partial Y^j} \frac{\partial Y^k}{\partial X^i} T^j_{(Y)} \frac{\partial}{\partial Y^k}$$

$$= \frac{\partial Y^k}{\partial Y^j} T^j_Y \frac{\partial}{\partial Y^k} = \delta^k_j T^j_Y \frac{\partial}{\partial Y^k} = T^j_{(Y)} \frac{\partial}{\partial Y^j} = T^i_{(Y)} \boldsymbol{e}_{i(Y)}. \quad (1.22)$$

We have therefore shown that the vector $\boldsymbol{T}$ is invariant under coordinate transformations, however, its components T^i do transform in a specific way. At this point the calculation (1.22) seems a little daunting, despite that it will soon become second nature.

Some of the older literature on differential geometry starts the entire subject with the definitions of transformation properties of vectors. Often the new coordinate system is denoted by a prime X'^i which simplifies the notation. A vector in the primed coordinate system also is indicated by a prime.

Definition 1.6 (Contravariant and covariant vectors). Let X^i be the coordinates of the vector space X and let X'^i be a different set of coordinates. A contravariant vector T^i transforms under coordinate transformation such that

$$T'^i = \frac{\partial X'^i}{\partial X^j} T^j. \quad (1.23)$$

A covariant vector S_i transforms under coordinate transformation such that

$$S'_i = \frac{\partial X^j}{\partial X'^i} S_j. \quad (1.24)$$

The covariant components of the vector S_i are defined with a lower index which indicates we should be able to write the vector $\boldsymbol{S}$ as follows:

$$\boldsymbol{S} = S_i \boldsymbol{E}^i, \quad (1.25)$$

where $\boldsymbol{E}^i$ are some new basis vectors with an upper index. Our next task is to interpret these basis vectors. Often the same letter is used for both types of basis vectors e_i and e^i where only the index position changes. For now it is slightly clearer to distinguish them.

Definition 1.7 (Dual vector space). Let V be a vector space. The dual vector space or dual space V^* is defined to the space of all linear mappings from V into $\mathbb{R}$. This means for $\boldsymbol{E} \in V^*$ and $e \in V$ we have $\boldsymbol{E} : V \mapsto \mathbb{R}$ and denote this as $\boldsymbol{E}(e)$ or $\boldsymbol{E} \cdot e$.

In Euclidean 3-space $V = \mathbb{E}^3$ we would identify standard column vectors with the elements of the space $\boldsymbol{v} \in V$ and row vectors with the elements of $\boldsymbol{v}^* \in V^*$. We would have

$$\boldsymbol{v}^*(\boldsymbol{v}) = \begin{pmatrix} v_1 & v_2 & v_3 \end{pmatrix} \begin{pmatrix} v^1 \\ v^1 \\ v^3 \end{pmatrix} = v_1 v^1 + v_2 v^2 + v_3 v^3 = v_i v^i. \quad (1.26)$$

We note that this is exactly the scalar product in Eq. (1.14). It is worth mentioning that the Einstein summation convention applies to one upper and one lower index which ensures that objects from the correct spaces are matched up to produce meaningful equations.

Note. For readers who have completed Quantum Mechanics, the Dirac ket vector $|\psi\rangle$ would naturally be associated with an element of V and the bra vector $\langle\phi|$ with an element of V^*. So $\langle\phi|$ is a mapping from V to the complex numbers in this case. $\langle\phi|\psi\rangle \in \mathbb{C}$ is the inner product between the two states. The spaces appearing in Quantum Mechanics are infinite dimensional and complex.

We recall the identification $e_i = \partial/\partial X^i$ where the e_i are the basis vectors of T_pV. Therefore the $\boldsymbol{E}^i$ are the basis vectors of $(T_pV)^*$. Let us denote these by $\boldsymbol{E}^i = dX^i$, at this stage this is nothing but a name, and we have

$$E^i(e_j) = dX^i \left(\frac{\partial}{\partial X^j} \right) = \delta^i_j, \tag{1.27}$$

meaning that the E^i and e_j are dual basis vector. One could equally write $E^i \cdot e_j = \delta^i_j$ as in Eq. (1.9). Next, we will show that one can indeed interpret the dual basis vectors as the coordinate differentials.

Let f be a smooth function and $\boldsymbol{T} = T^i e_i$ be an element of $T_p V$. Let us start with the usual classical differential

$$df = \frac{\partial f}{\partial X^i} dX^i. \tag{1.28}$$

According to our naming, this object should be an element of $(T_p V)^*$. This all works out nicely when we consider

$$df(\boldsymbol{T}) = \frac{\partial f}{\partial X^i} dX^i (T^j e_j) = \frac{\partial f}{\partial X^i} T^j E^i e_j$$

$$= \frac{\partial f}{\partial X^i} T^j \delta^i_j = T^i \frac{\partial f}{\partial X^i}. \tag{1.29}$$

This means $df(\boldsymbol{T})$ is nothing but the directional derivative of f along the vector $\boldsymbol{T}$. This is a very nice result which connects vector calculus with geometry, hence the name differential geometry.

We will now introduce the comma notation whereby partial derivatives with respect to the coordinate X^i are denoted by $_{,i}$ or by ∂_i, this means we define

$$f_{,i} := \frac{\partial f}{\partial X^i} =: \partial_i f, \quad \text{and} \quad f_{,ij} := \frac{\partial^2 f}{\partial X^i \partial X^j} =: \partial_{ij} f. \tag{1.30}$$

This is very convenient, in particular in vector calculus. In this case we are working in $\mathbb{E}^3$ and have

$$\operatorname{grad} f = \nabla f = f_{,i} e^i, \tag{1.31}$$

$$\operatorname{div} \boldsymbol{A} = \nabla \cdot \boldsymbol{A} = A^i{}_{,i}, \tag{1.32}$$

$$\operatorname{curl} \boldsymbol{A} = \nabla \times \boldsymbol{A} = \varepsilon^{ijk} \partial_i A_j e_k. \tag{1.33}$$

The components of df in Eq. (1.28) are given by grad f. The curl operator is specific to three (and seven) dimensions and does not exist in this form in four dimensions for instance. Many vector calculus identities can be proved very efficiently using the index notation. In vector calculus, in three dimensions, it is common to use the symbol ∇, however, in differential geometry this is reserved for the covariant derivative.

1.2.2.   *Tensors*

The definitions of contravariant and covariant vectors can be extended to objects with more than one index. The motivation for this stems from the fact that some objects in nature simply cannot be written as scalar or vector quantities. One such example is the Cauchy stress tensor which is a rank-2 tensor. The rank of a tensor simply counts its number of indices. Another example is the permittivity tensor in electromagnetism. The permittivity tensor relates the exterior electric field E to the electric displacement field D in the medium. In linear elasticity theory the stresses and strains are related by the material tensor, which is a rank-4 tensor, it relates two rank-2 objects. In simple words, a tensor is something where every index transforms correctly under coordinate transformations. More precisely this is written as follows.

Definition 1.8 (Tensor of type (p,q)). A tensor of type (p, q) is an object with p contravariant and q covariant indices, so we write $T^{i_1 i_2 \cdots i_p}{}_{j_1 j_2 \cdots j_q}$. Under coordinate transformations it transforms according to

$$T'^{i_1 i_2 \cdots i_p}{}_{j_1 j_2 \cdots j_q} = \frac{\partial X'^{i_1}}{\partial X^{k_1}} \cdots \frac{\partial X'^{i_p}}{\partial X^{k_p}} \frac{\partial X^{m_1}}{\partial X'^{j_1}} \cdots \frac{\partial X^{m_q}}{\partial X'^{j_q}} T'^{k_1 k_2 \cdots k_p}{}_{m_1 m_2 \cdots m_q}.$$
$$(1.34)$$

The rank of a tensor is the number of its indices, $r = p + q$.

Tensors of the same type can be added and subtracted, and we can multiply any tensor by an arbitrary (real) scalar. For tensors with rank $r \geq 2$, there is an additional operation, namely the contraction.

Definition 1.9 (Contraction). Let $T^{i_1 i_2 \cdots i_p}{}_{j_1 j_2 \cdots j_q}$ be a tensor of type (p, q). The object

$$T^{i_1 i_2 \cdots k \cdots i_p}{}_{j_1 j_2 \cdots k \cdots j_q}, \tag{1.35}$$

is a type $(p - 1, q - 1)$ tensor with rank $p + q - 2$. We say that we have contracted over one pair of indices. Note that we sum over the index k.

For a rank-2 tensor, the contraction corresponds to taking the trace of a square matrix, A for instance. We have $\operatorname{tr} A = A^i{}_i$ which is the sum of the diagonal elements. One can think of the contraction as a higher dimensional version of the trace. For instance, if we begin with a rank $2n$ tensor, then we can in principle contract n-times and arrive at a scalar. For a tensor of rank $2n + 1$ we can also contract n times, however, in this case one arrives at a vector.

Let T_{ab} be a rank-2 tensor, then we define the symmetric and skew-symmetric parts as follows:

$$T_{(ab)} = \frac{1}{2}(T_{ab} + T_{ba}), \tag{1.36}$$

$$T_{[ab]} = \frac{1}{2}(T_{ab} - T_{ba}). \tag{1.37}$$

Every tensor can be written as $T_{ab} = T_{(ab)} + T_{[ab]}$. A rank-2 tensor is called symmetric if $T_{ab} = T_{ba}$ and skew-symmetric if $T_{ab} = -T_{ba}$. This is in complete analogy to matrices. Let A be a square matrix, then one defines the symmetric part by $\operatorname{sym}(A) = (A + A^T)/2$ and its skew-symmetric part by $\operatorname{skew}(A) = (A - A^T)/2$. It is natural to view rank-2 tensors as square matrices, however, one has to be very careful since tensor components transform under coordinate transformations. Note that one can generalise Eqs. (1.36) and (1.37) to higher rank tensors, however, we will not use this notation as it is easier to read equations written out explicitly, for the time being.

In n dimensions any symmetric rank-2 tensor has $n(n+1)/2$ independent components, and any skew-symmetric rank-2 tensor has $(n-1)n/2$ independent components. We note that these two numbers add to n^2 as expected.

We have seen that $\partial f / \partial X^i$ transforms like a covariant vector under coordinate transformations. Let us check how that partial derivative of a contravariant vector A^i transforms under coordinate transformations

$$
\begin{aligned}
\frac{\partial A'^i}{\partial X'^j} &= \frac{\partial X^k}{\partial X'^j} \frac{\partial}{\partial X^k} \left(\frac{\partial X'^i}{\partial X^m} A^m \right) \\
&= \frac{\partial X^k}{\partial X'^j} \left(\frac{\partial^2 X'^i}{\partial X^m \partial X^k} A^m + \frac{\partial X'^i}{\partial X^m} \frac{\partial A^m}{\partial X^k} \right) \\
&= \frac{\partial X^k}{\partial X'^j} \frac{\partial X'^i}{\partial X^m} \frac{\partial A^m}{\partial X^k} + \frac{\partial X^k}{\partial X'^j} \frac{\partial^2 X'^i}{\partial X^m \partial X^k} A^m.
\end{aligned} \tag{1.38}
$$

The first term on this last line transforms correctly, however, we have an additional second term. Therefore, the object A^i_j is not a $(1,1)$ tensor and hence the partial derivative is not a 'good' derivative operator. We need to define a new derivative operator which maps tensors to tensors.

However, we can be clever and consider the object F^i_j given by

$$
F^i_j = \partial_i A^j - \partial^j A_j. \tag{1.39}
$$

We can view this as the skew-symmetric part of this partial derivative without the factor of $1/2$, or the Faraday tensor of electromagnetism. If we swap the indices i and j in Eq. (1.38) and then compute the transformation properties of F^i_j we note that the terms which do not transform correctly cancel each other. Thus, F^i_j does transform like a $(1,1)$ tensor which is a nice little result since we have not yet defined a meaningful derivative operator.

1.2.3. *Manifolds and metric*

Before defining what is meant by a manifold, let us briefly discuss the 2-sphere $\mathbb{S}^2$ again. Normally, when we think of this space we view it embedded in $\mathbb{E}^3$. However, the geometrical properties of the sphere should really be independent of this embedding. The idea of the manifold is to make this more precise.

Definition 1.10 (Manifold). An n-dimensional smooth manifold $\mathscr{M}$ is a set together with a collection of open subsets $\{\Omega_\alpha\}$ and a collection of coordinates $\{X^i_{(\alpha)}\}$ with $\alpha \in \mathbb{N}$ and $i = 1, \ldots, n$. These satisfy:

(i) Every point $p \in \mathscr{M}$ is also in at least one $\{\Omega_\alpha\}$. This means that the collection $\{\Omega_\alpha\}$ will contain every point p, we say that the manifold $\mathscr{M}$ is covered by $\{\Omega_\alpha\}$.

(ii) For each α the coordinates $\{X^i_{(\alpha)}\}$ are bijective functions $X^i_{(\alpha)} : \Omega_\alpha \mapsto O_\alpha \subset \mathbb{R}^n$ where the O_α are also open.

(iii) If any U_α and U_β overlap, then the coordinates $\{X^i_{(\alpha)}\}$ and $\{X^i_{(\beta)}\}$ are related by a coordinate transformation on the intersection $U_\alpha \cap U_\beta$.

A manifold is something very natural in a way. It means we have some arbitrarily shaped space we are trying to understand, however, we cannot do this directly. So, we find some open subsets with coordinates which cover this space using one patch at a time. These patches all come with some coordinates which then allows us to locally understand this particular patch. Whenever two such patches overlap, we can transform coordinates from this patch to the next and carry on. In this way we are able to study the entire manifold. Physicists tend to call the $\{X^i_{(\alpha)}\}$ for a specific α a coordinate system while mathematicians prefer the word chart. One can also define complex manifolds by working with $\mathbb{C}^n$ instead of $\mathbb{R}^n$. Figure 1.2 is useful in visualising the idea of a manifold.

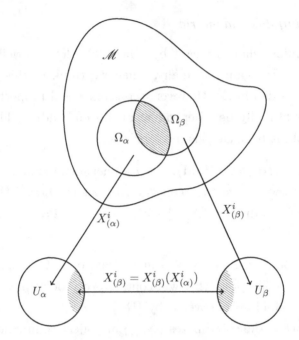

Fig. 1.2   Manifold $\mathcal{M}$ covered by two coordinate two patches Ω_α and Ω_β. The intersection $\Omega_\alpha \cap \Omega_\beta$ is indicated by the shaded regions. There exists a coordinate transformation between highlighted parts of U_α and U_β.

Many spaces can be described using the concept of a manifold, examples are flat spaces in n dimensions, the n-spheres $\mathbb{S}^n$ and n-tori. Other examples are the Möbius strip and Klein's bottle.

Having defined a manifold on which we can define scalars, vectors and tensors, we must now add some additional structures. In the end we want to build a physical theory which means we want to be able to measure distances between points, lengths of vectors, and also angles between curves for instance. This additional structure is provided by the metric g_{ab} which is a symmetric rank-2 tensor with inverse g^{bc} so that $g_{ab}g^{bc} = \delta_a^c$. One way to motivate the metric is as follows. Let us start with Pythagoras' theorem. If the point P has coordinate (x, y, z) with respect to the origin, then the distance of P from that origin is given by $s^2 = x^2 + y^2 + z^2$. Next, we consider small values

Δx, Δy, ... and then pass to the infinitesimal limit and we write

$$ds^2 = dx^2 + dy^2 + dz^2. \tag{1.40}$$

This says that the infinitesimal distances between points is position independent. This space is the same everywhere. When we introduced vectors earlier we discussed the sphere $\mathbb{S}^2$ and noted the difficulties of defining vectors on the surface of the sphere. In order to define the 'correct' distance between two points on the sphere, we cannot simply use Pythagoras' theorem as this line connecting the two points would go through part of the sphere and not be on its surface. In order to be able to measure distances correctly for all kinds of spaces, we need to generalise Eq. (1.40). Therefore, we need to introduce arbitrary functions in front of all the squared infinitesimal and all the possible cross terms. This means we will write

$$ds^2 = g_{ij}\, dX^i dX^j, \tag{1.41}$$

where all the g_{ij} are functions of the coordinates. The object g_{ij} is called the metric tensor, often just called the metric. The object ds^2 is called the line element, it is often used as a synonym for metric. Since $dX^i dX^j$ is symmetric in i and j, g_{ij} must also be symmetric. In 3 dimensions the metric contains 6 functions, in 4 dimensions it contains 10 functions, and in dimension n there are $n(n+1)/2$ functions in the metric.

Another way to motivate (define) the metric is by going back to an arbitrary vector $\boldsymbol{V}$ which we could write in two different ways $\boldsymbol{V} = V_i \boldsymbol{E}^i = V^i \boldsymbol{e}_i$. So far we avoided the question of how the components V_i are related to those of V^i. We can define g_{ij} to be the mapping which relates these quantities by requiring $X_i = g_{ij} X^j$. This corresponds to viewing the metric as the inner product of this space. However, this will only work if the metric is positive definite and it turns out that in General Relativity our metric tensor is not positive definite. We will put this into the following definition.

Definition 1.11 (Metric and line element). Let $\mathcal{M}$ be an n-dimensional smooth manifold. The Riemannian metric tensor g_{ij} defines the positive definite inner product $\langle\,,\rangle$ such that

$$\langle U, V \rangle = g_{ij}U^iV^j. \tag{1.42}$$

If $\langle\,,\rangle$ is only non-degenerate,[1] then we call the corresponding metric pseudo-Riemannian. The line element ds^2 is

$$ds^2 = g_{ij}\, dX^i dX^j, \tag{1.43}$$

where the X^i are the coordinates in a patch of $\mathcal{M}$.

If we evaluate the metric tensor at some point on the manifold, then, being a symmetric $n \times n$ matrix, it will have n real eigenvalues, taking into account multiplicity. The number of positive, negative or zero eigenvalues cannot be changed by coordinate transformations.

Definition 1.12 (Metric signature). The signature of the metric is the number of positive, negative and zero eigenvalues, however, we will not encounter a metric with zero eigenvalues. In general, the pair (p, q) of integers denotes the number of positive and negative eigenvalues, respectively. Sometimes the signs of the eigenvalues are made explicit, instead of writing $(3, 1)$ one often writes the signature is $(+, +, +, -)$ or $(-, +, +, +)$ as the order can be changed by relabelling coordinates.

Definition 1.13 (Lorentzian metric). A metric is called Lorentzian if it has signature $(p, 1)$ or $(1, q)$.

The Lorentzian metric is the fundamental variable of General Relativity. The theory is formulated in four dimensions, one time direction and three space directions, and the Einstein field equations are a set of 10 nonlinear partial differential equations in the

---

[1] This means $\langle U, V \rangle = 0$ for all V implies that $U = 0$.

components of the metric tensor. Solving the Einstein field equations means finding the metric functions.

1.2.4. *Examples of metrics*

This section will contain some examples of the Euclidean metric in two and three dimensions using different common coordinate systems. The reader is encouraged to work through these examples as they will be used at various points in the remainder of the book. The standard line elements of Euclidean space using Cartesian coordinates are

$$ds^2 = dx^2 + dy^2, \tag{1.44}$$

$$ds^2 = dx^2 + dy^2 + dz^2, \tag{1.45}$$

in two and three dimensions, respectively.

Example 1.2 (Polar coordinates in two-dimensional (2D)).
Let us introduce $x = r \cos \varphi$ and $y = r \sin \varphi$, then

$$dx = \cos \varphi dr - r \sin \varphi d\varphi, \tag{1.46}$$

$$dy = \sin \varphi dr + r \cos \varphi d\varphi, \tag{1.47}$$

$$dx^2 = \cos^2 \varphi dr^2 + r^2 \sin^2 \varphi d\varphi^2 - 2r \sin \varphi \cos \varphi dr d\varphi, \tag{1.48}$$

$$dy^2 = \sin^2 \varphi dr^2 + r^2 \cos^2 \varphi d\varphi^2 + 2r \sin \varphi \cos \varphi dr d\varphi. \tag{1.49}$$

Adding up the last two equations shows that Euclidean space in polar coordinates is given by

$$ds^2 = dx^2 + dy^2 = dr^2 + r^2 d\varphi^2. \tag{1.50}$$

Example 1.3 (Cylindrical coordinates in 3D). We introduce $x = \rho \cos \varphi$, $y = \rho \sin \varphi$ and $z = z$. We can follow the previous calculation and arrive at the metric of Euclidean space in cylindrical coordinates

$$ds^2 = dx^2 + dy^2 + dz^2 = d\rho^2 + \rho^2 d\varphi^2 + dz^2. \qquad (1.51)$$

The use of ρ instead of r is useful as the letter r is generally associated with the Euclidean distance from the origin. In this case $r^2 = \rho^2 + z^2$.

Example 1.4 (Spherical coordinates in 3D). Our conventions for the spherical coordinate system are

$$x = r\sin\theta\cos\phi, \qquad (1.52)$$
$$y = r\sin\theta\sin\phi, \qquad (1.53)$$
$$z = r\cos\theta. \qquad (1.54)$$

Computing all the terms dx, dx^2, dy, ... is slightly lengthy but straightforward. The result is

$$ds^2 = dx^2 + dy^2 + dz^2 = dr^2 + r^2 d\theta^2 + r^2 \sin^2\theta d\phi^2. \qquad (1.55)$$

Therefore, we can write the metric tensor as

$$g_{ij} = \begin{pmatrix} 1 & 0 & 0 \\ 0 & r^2 & 0 \\ 0 & 0 & r^2\sin^2\theta \end{pmatrix}. \qquad (1.56)$$

The determinant of this metric tensor $g = \det g_{ij}$ is the product of its diagonal components, and $\sqrt{g} = r^2\sin\theta$. This is the Jacobian of the transformation from Cartesian to spherical coordinates which is well known in vector calculus.

Following on from the previous examples, without providing further details at this point, we state that the volume of a space with given metric tensor g_{ij} is given by

$$V = \int \sqrt{|\det g_{ij}|}\, d^n x. \qquad (1.57)$$

Example 1.5 (Surface of the 2-sphere). Continuing on from the previous example, we set $r = R = $ const. in order to describe the surface of the sphere. The corresponding line element now becomes

$$ds^2 = R^2 \left(d\theta^2 + \sin^2 \theta d\phi^2 \right).$$ (1.58)

It is customary in General Relativity to introduce the notation $d\Omega^2 = d\theta^2 + \sin^2 \theta d\phi^2$ where $d\Omega^2$ is the line element of the unit sphere.

Example 1.6 (Poincaré hyperbolic disk). The line element of the Poincaré hyperbolic disk is given by

$$ds^2 = 4 \frac{dx^2 + dy^2}{(1 - x^2 - y^2)^2},$$ (1.59)

with $x^2 + y^2 < 1$. This is a really interesting space which inspired some of the works of Escher, Circle Limit III (1959) and Circle Limit IV (1960), see Escher (2015). This is an example of a space with constant negative curvature, we will define later what this means. Such spaces are important in Cosmology where we will meet them again.

Example 1.7 (Minkowski space). The line element of Minkowski space is

$$ds^2 = -c^2 dt^2 + dx^2 + dy^2 + dz^2.$$ (1.60)

This is our first example of a Lorentzian metric with signature $(-, +, +, +)$, here c is the speed of light. This is required because we need the physical units to match. Alternatively, one could have divided the spatial parts by c^2 instead of multiplying the time component. However, conventionally the line element has units of length. The metric functions are dimensionless.

It is fair to say that Minkowski space is the most important space of theoretical physics. Consequently, the Minkowski metric is denoted by a separate symbol, namely η_{ij}.

1.2.5. *Geodesics*

Let us consider a manifold $\mathcal{M}$ and a curve C given by $X^i = X^i(\tau)$ with parameter τ. Let us simply start with a 2D flat space with line element $ds^2 = dx^2 + dy^2$ and coordinates $X^i = (x, y)$. Our curve C is now given in parametric form $(x, y) = (x(\tau), y(\tau))$. Let us now eliminate that parameter in the parametric form and write our curve as $y = y(x)$, just like in calculus. We will now show that ds, the square root of the line element, corresponds to the arc length of this curve. Since $y = y(x)$, we have $dy = y'(x)dx$ and therefore $ds^2 = dx^2 + y'(x)^2 dx^2 = (1 + y'(x)^2)dx^2$. Now we can write

$$s = \int ds = \int \sqrt{1 + y'(x)^2} dx, \tag{1.61}$$

which is the well-known calculus formula.

Consider an arbitrary curve C given by $X^i = X^i(\lambda)$, where λ is the affine parameter, see Definition 1.3. In a space with metric g_{ij} we have

$$s = \int ds = \int \sqrt{g_{ij} \dot{X}^i \dot{X}^j} d\lambda, \tag{1.62}$$

where the dot means differentiation with respect to λ.

This means, if we have two points which are connected by C, then we can use the arc length s to determine the distance between these points along that curve. However, we are much more interested in special curves, namely those that minimise the distance between our points. In other words, what is the shortest distance? From a physical point of view this is in fact the most natural question, related to the principle of least action (Hamilton's principle) which is fundamental to modern physics.

In order to find the shortest lines between any two given points, we treat Eq. (1.62) as the action functional and the integrand as the

Lagrangian function of the system.[2] Therefore, we write

$$L(X^i, \dot{X}^j) = \sqrt{g_{ij}\dot{X}^i\dot{X}^j}, \qquad (1.63)$$

where we must remember that the metric is a function of the coordinates X^i. Note that our Lagrangian does not depend on λ explicitly. The Euler–Lagrange equations are given by

$$\frac{\partial L}{\partial X^k} = \frac{d}{d\lambda}\frac{\partial L}{\partial \dot{X}^k}. \qquad (1.64)$$

Since we work in n dimensions, we will find n equations, labelled by the index k. The left-hand side of Eq. (1.64) is easily computed

$$\frac{\partial L}{\partial X^k} = \frac{1}{2L}\frac{\partial g_{ij}}{\partial X^k}\dot{X}^i\dot{X}^j. \qquad (1.65)$$

We also find

$$\frac{\partial L}{\partial \dot{X}^k} = \frac{1}{2L}g_{ij}\left(\delta_k^i\dot{X}^j + \dot{X}^i\delta_k^j\right) = \frac{1}{L}g_{ki}\dot{X}^i. \qquad (1.66)$$

The term in the brackets is the result of applying the chain rule. In order to find the right-hand side of Eq. (1.64), we need to differentiate the latter with respect to λ. Remembering that our parametrisation is affine we have $L = 1$ and hence

$$\frac{d}{d\lambda}\frac{\partial L}{\partial \dot{X}^k} = \frac{\partial g_{ki}}{\partial X^m}\dot{X}^m\dot{X}^i + g_{ki}\ddot{X}^i. \qquad (1.67)$$

Consequently the complete Euler–Lagrange equations are given by

$$\frac{1}{2}\frac{\partial g_{ij}}{\partial X^k}\dot{X}^i\dot{X}^j = \frac{\partial g_{ki}}{\partial X^m}\dot{X}^m\dot{X}^i + g_{ki}\ddot{X}^i. \qquad (1.68)$$

Next, we will rewrite these equations by introducing a new symbol, called the Christoffel symbol, which plays a paramount role in

---

[2]For readers unfamiliar with the concepts of variations and Euler–Lagrange equations, a very brief introduction is given in Sec. 1.4.

Differential Geometry and General Relativity. We start with

$$g_{ki}\ddot{X}^i + \frac{1}{2}(\partial_m g_{ki}\dot{X}^m\dot{X}^i + \partial_m g_{ki}\dot{X}^m\dot{X}^i - \partial_k g_{ij}\dot{X}^i\dot{X}^j) = 0 \quad (1.69)$$

and apply g^{nk} to this equation and arrive at

$$\ddot{X}^n + \Gamma^n_{ij}\dot{X}^i\dot{X}^j = 0, \quad (1.70)$$

where the Christoffel symbol or connection Γ^n_{ij} is given by

$$\Gamma^n_{ij} = \frac{1}{2}g^{nk}\left(\partial_i g_{jk} + \partial_j g_{ki} - \partial_k g_{ij}\right). \quad (1.71)$$

This object is symmetric in the lower pair of indices. Equation (1.70) is the so-called geodesic equation. Sometimes the Christoffel symbol is denoted differently, using curly brackets

$$\left\{{}^n_{ij}\right\} = \frac{1}{2}g^{nk}\left(\partial_i g_{jk} + \partial_j g_{ki} - \partial_k g_{ij}\right), \quad (1.72)$$

which is useful if one needs to distinguish the Christoffel symbol from a different connection Γ^n_{ij}. Note that older textbooks tend to use the curly bracket notation. Let us summarise this result as a definition.

Definition 1.14 (Geodesics). A curve C given by $X^i(\tau)$ with affine parameter τ is called a geodesic if it satisfies the equation

$$\ddot{X}^n + \Gamma^n_{ij}\dot{X}^i\dot{X}^j = 0, \quad (1.73)$$

where the Γ^n_{ij} are given by Eq. (1.71).

Every curve C which satisfies this equation connects points via shortest distances. As expected, this depends on the metric functions. In 3D Euclidean space the metric looks like the identity matrix, in other words the line element is $ds^2 = dx^2 + dy^2 + dz^2$. Hence, all partial derivatives of the metric vanish identically and the geodesic equation reduces to $\ddot{X}^n = 0$ which we can integrate twice. We find that straight lines are the shortest lines in flat space, as one would expect. The geodesic equations also contain the first clue towards a geometrical theory of gravity which we will discuss in Sec. 2.2.

By definition, the Christoffel symbol is symmetric in its lower pair of indices. Therefore, this pair gives rise to $n(n+1)/2$ components in n dimensions. There are no other symmetries, so the third index can take any value and hence the Christoffel symbol has $n^2(n+1)/2$ independent components in n dimensions. In dimensions $2, 3, 4$ the Christoffel symbol has $6, 18, 40$ components, respectively.

For a curve with affine parametrisation, the geodesic equations can also be derived from considering the simpler Lagrangian

$$L(X^i, \dot{X}^j) = g_{ij} \dot{X}^i \dot{X}^j, \tag{1.74}$$

this is Eq. (1.63) without the square root. The advantage of this formulation is that it provides us with an efficient tool to compute the components of the Christoffel symbol. We will work through some examples next which show the two different ways of finding the Christoffel symbols and the geodesic equations.

Example 1.8 (Euclidean space with polar coordinates). We choose coordinates $X^1 = r$ and $X^2 = \varphi$, the line element of this space is given by $ds^2 = dr^2 + r^2 d\varphi^2$, so that the metric and the inverse metric are

$$g_{ab} = \begin{pmatrix} 1 & 0 \\ 0 & r^2 \end{pmatrix}, \quad g^{ab} = \begin{pmatrix} 1 & 0 \\ 0 & r^{-2} \end{pmatrix}. \tag{1.75}$$

Let us compute the Christoffel symbol components which are defined by Eq. (1.71).

$$\Gamma^1_{ij} = \frac{1}{2} g^{1k} \left(\partial_i g_{jk} + \partial_j g_{ki} - \partial_k g_{ij} \right). \tag{1.76}$$

Since the inverse metric is diagonal, the only contribution from g^{1k} (remember we are summing over k) comes from $k = 1$, therefore

$$\Gamma^1_{ij} = \frac{1}{2} g^{11} \left(\partial_i g_{j1} + \partial_j g_{1i} - \partial_1 g_{ij} \right). \tag{1.77}$$

Also the metric is diagonal which simplifies the calculation. We find

$$\Gamma^1_{11} = 0, \quad \Gamma^1_{12} = 0, \quad \Gamma^1_{22} = -r, \tag{1.78}$$

which follows directly from the previous equation. Now we choose $k = 2$

$$\Gamma_{ij}^2 = \frac{1}{2}g^{22}\left(\partial_i g_{j2} + \partial_j g_{2i} - \partial_2 g_{ij}\right). \tag{1.79}$$

When $i = j = 1$ or $i = j = 2$, all terms vanish. However, if we choose $i = 1$ and $j = 2$ we find

$$\Gamma_{12}^2 = \frac{1}{2}g^{22}\partial_1 g_{22} = \frac{1}{2}\frac{1}{r^2}2r = \frac{1}{r}, \tag{1.80}$$

therefore, the other components are given by

$$\Gamma_{11}^2 = 0, \quad \Gamma_{12}^2 = \frac{1}{r}, \quad \Gamma_{22}^2 = 0. \tag{1.81}$$

We found all six components.

Example 1.9 (Euclidean space with polar coordinates again). Now we show the second method of finding the Christoffel symbol components. We start with $L = g_{ij}\dot{X}^i\dot{X}^j$ which for polar coordinates and metric Eq. (1.75) becomes

$$L = \dot{r}^2 + r^2\dot{\varphi}^2. \tag{1.82}$$

The two Euler–Lagrange equations of this Lagrangian are given by

$$\frac{\partial L}{\partial r} = \frac{d}{d\lambda}\frac{\partial L}{\partial \dot{r}} \quad \Rightarrow \quad 2r\dot{\varphi}^2 = \frac{d}{d\lambda}(2\dot{r}), \tag{1.83}$$

$$\frac{\partial L}{\partial \varphi} = \frac{d}{d\lambda}\frac{\partial L}{\partial \dot{\varphi}} \quad \Rightarrow \quad 0 = \frac{d}{d\lambda}(2r^2\dot{\varphi}). \tag{1.84}$$

By applying the product rule to the right-hand sides, we arrive at these two equations

$$\ddot{r} - r\dot{\varphi}^2 = 0, \tag{1.85}$$

$$\ddot{\varphi} + 2\frac{1}{r}\dot{r}\dot{\varphi} = 0. \tag{1.86}$$

By virtue of the geodesic equations, we can directly read of the Christoffel symbol components. From the $\ddot{r}$ equations we find

$$\Gamma_{11}^1 = 0, \quad \Gamma_{12}^1 = 0, \quad \Gamma_{22}^1 = -r, \tag{1.87}$$

while the $\ddot{\varphi}$ equation results in

$$\Gamma^2_{11} = 0, \quad \Gamma^2_{12} = \frac{1}{r}, \quad \Gamma^2_{22} = 0. \tag{1.88}$$

This is in agreement with the previous example. However, this approach tends to be more efficient for computing these components.

Example 1.10. Show that the geodesic equations based on the line element $ds^2 = dr^2 + r^2 d\varphi^2$ correspond to straight lines. Following on from the previous example, the geodesic equations are given by

$$\ddot{r} - r\dot{\varphi}^2 = 0, \tag{1.89}$$

$$\frac{d}{d\lambda}(2r^2\dot{\varphi}) = 0. \tag{1.90}$$

Consequently, we can integrate the second equation and find $2r^2\dot{\varphi} = 2\ell$ where ℓ is some constant of integration. This means that the angular velocity of geodesic curves is constant, or more physically speaking, angular momentum is conserved. We can now replace the $\dot{\varphi}$ term in Eq. (1.89) and find

$$\ddot{r} = \frac{\ell^2}{r^3}. \tag{1.91}$$

One can solve this equation as follows. First, we multiply by $\dot{r}$ and integrate which yields

$$\dot{r}^2 = C_1 - \frac{\ell^2}{r^2}, \tag{1.92}$$

where C_1 is a constant of integration. Next, separation of variables and integration leads to

$$r(\lambda) = \sqrt{\frac{\ell^2}{C_1} + C_1(\lambda + C_2)^2}, \tag{1.93}$$

where C_2 is another constant of integration. Lastly, we can find $\varphi(\lambda)$ by integrating $\dot{\varphi} = \ell/r^2$ which gives

$$\varphi(\lambda) = \arctan\left(\frac{C_1}{\sqrt{\ell}}(\lambda + C_2)\right) + \arctan(C_3), \tag{1.94}$$

where C_3 is yet another constant of integration. In total we have four constants of integration. This is consistent with having two second-order equations to solve. It remains to show that the system of Eqs. (1.92) and (1.93) are indeed straight lines in polar coordinates. For this we note that

$$r\cos(\varphi) = \frac{\ell - C_1 C_3 (C_2 + \lambda)}{\sqrt{C_1}\sqrt{1 + C_3^2}}, \qquad (1.95)$$

$$r\sin(\varphi) = \frac{\ell C_3 + C_1 (C_2 + \lambda)}{\sqrt{C_1}\sqrt{1 + C_3^2}}. \qquad (1.96)$$

Therefore, we can write

$$r\sin(\varphi) = -\frac{1}{C_3} r\cos(\varphi) + \frac{\ell\sqrt{1 + C_3^2}}{C_3\sqrt{C_1}}, \qquad (1.97)$$

which looks like the equation of a straight line $y = kx + m$ by choosing x, y and the constants appropriately.

This calculation was rather painful and would have been rather trivial if we worked in appropriate coordinates. There are two things we should take away from this example: First, whenever we are interested in performing explicit calculations we should ensure that our choice of coordinates is as smart as possible. Second, there are many long calculations in differential geometry which cannot be avoided, irrespective of coordinates, so it is a good idea to practise them even if this means calculating straight lines in a rather unpleasant way.

In cases when we need to compute the trace of the Christoffel symbol Γ_{ab}^b, there exists a useful and simple formula based on the determinant of the metric tensor $g = \det(g_{ij})$. We start recalling the well-known formula

$$\frac{d(\det \mathbf{M})}{dx} = \det \mathbf{M} \, \text{tr} \left[\mathbf{M}^{-1} \frac{d\mathbf{M}}{dx} \right], \qquad (1.98)$$

which we now rewrite in index notation applied to the metric tensor

$$\partial_i g = g \, g^{ab} \partial_i g_{ba}. \tag{1.99}$$

Going back to Eq. (1.71) and summing over n and j we find

$$\Gamma^j_{ij} = \frac{1}{2} g^{jk} \partial_i g_{jk} = \frac{1}{2} \frac{1}{g} \partial_i g = \frac{1}{\sqrt{|g|}} \partial_i \sqrt{|g|} = \partial_i \left(\log \sqrt{|g|} \right). \tag{1.100}$$

Example 1.11 (Euclidean space with polar coordinates). We already computed the Christoffel symbol components, see Eqs. (1.80) and (1.81). Consequently,

$$\Gamma^j_{1j} = \Gamma^1_{11} + \Gamma^2_{12} = \frac{1}{r}, \quad \Gamma^j_{2j} = \Gamma^1_{21} + \Gamma^2_{22} = 0. \tag{1.101}$$

Now let us check this result using Eq. (1.100). The determinant of the metric is $g = r^2$ and hence $\sqrt{|g|} = r$. Therefore, $\partial_\varphi (\log r) = 0$, and also $\partial_r (\log r) = 1/r$, both as expected.

1.2.6. *Covariant derivative*

Since the Christoffel symbol depends on the first partial derivatives of the metric tensor, it will not transform like a tensor, recall Eq. (1.38). However, this might be quite useful: Can we combine the partial derivative of a vector A^i with the Christoffel symbol so that this new object transforms correctly under coordinate transformations?

To answer this question, let us find the transformation properties of $\Gamma^n_{ij} A^j$. A direct calculation shows that

$$\Gamma'^n_{ij} = \frac{\partial X'^n}{\partial X^a} \frac{\partial X^b}{\partial X'^i} \frac{\partial X^c}{\partial X'^j} \Gamma^a_{bc} + \frac{\partial^2 X^m}{\partial X'^i \partial X'^j} \frac{\partial X'^n}{\partial X^m}, \tag{1.102}$$

from which we can compute the transformation properties of $\Gamma'^n_{ij} A^j$, see Exercise 1.9. One observes that the inhomogeneous terms, those that do not transform correctly, are the same as those of $\partial_i A^n$ with a different sign. Therefore, we can use the partial derivative and the Christoffel symbol to create a new derivative operator which would transform like a tensor under coordinate transformation.

Before defining the so-called covariant derivative, we introduce this more formally and then show that we arrive at the same conclusions despite following a rather different route.

Definition 1.15 (Covariant derivative). A covariant derivative (sometimes derivative operator) ∇_a on a manifold $\mathcal{M}$ is a mapping which takes a type (p, q) tensor to a tensor of type $(p, q+1)$ with the following properties:

(i) For any smooth function f the covariant derivative coincides with the partial derivative

$$\nabla_a f = \partial_a f. \tag{1.103}$$

(ii) The derivative is linear, this means for all $\alpha, \beta \in \mathbb{R}$

$$\nabla_a \left(\alpha A^{\cdots}{}_{\cdots} + \beta B^{\cdots}{}_{\cdots} \right) = \alpha \nabla_a A^{\cdots}{}_{\cdots} + \beta \nabla_a B^{\cdots}{}_{\cdots}, \tag{1.104}$$

where the dots indicate tensors of arbitrary rank of the same type.

(iii) The derivative satisfies the Leibniz rule (or product rule)

$$\nabla_a \left(A^{\cdots}{}_{\cdots} B^{\cdots}{}_{\cdots} \right) = \left(\nabla_a A^{\cdots}{}_{\cdots} \right) B^{\cdots}{}_{\cdots} + A^{\cdots}{}_{\cdots} \left(\nabla_a B^{\cdots}{}_{\cdots} \right). \tag{1.105}$$

(iv) The derivative commutes with contraction

$$\nabla_a A^{a_1 \cdots k \cdots a_m}{}_{b_1 \cdots k \cdots b_n} = \nabla_a \left(\delta^k_j A^{a_1 \cdots j \cdots a_m}{}_{b_1 \cdots k \cdots b_n} \right). \tag{1.106}$$

Let us briefly discuss the implications of the fourth property. For the derivative to commute with the contraction is equivalent to the requirement $\nabla_a \delta^k_j = 0$. Recalling that $g^{km} g_{jm} = \delta^k_j$ means that

$$0 = \nabla_a \left(g^{km} g_{jm} \right) = (\nabla_a g^{km}) g_{jm} + g^{km} \nabla_a g_{jm}, \tag{1.107}$$

or equivalently $\nabla_a g^{ki} = -g^{ij} g^{km} \nabla_a g_{jm}$. The interesting point is that there are two possibilities of satisfying this relationship. The simplest solution is to require that the covariant derivative of the metric tensor vanishes. This leads us to the next definition.

Definition 1.16 (Metricity and non-metricity). A covariant derivative is called metric compatible or simply metric if it satisfies $\nabla_a g_{ij} = 0$ and non-metric otherwise. The object of non-metricity is defined by $\nabla_a g_{ij} = Q_{aij}$, it is a rank-3 tensor symmetric in the last two indices.

Next, let us have a closer look at the first property of the covariant derivative which states that $\nabla_a f = \partial_a f$. Let us take this relation and differentiate again covariantly. This gives rise to the object $\nabla_b \nabla_a f = \nabla_b \partial_a f$. Had we started with the indices the other way round, we would have arrived at $\nabla_a \nabla_b f = \nabla_a \partial_b f$ and there is no *a priori*[3] reason why these two should be the same. Hence, covariant derivatives of scalars and tensors in general do not commute which motivates the next definition.

Definition 1.17 (Torsion). A covariant derivative is called torsion-free if it satisfies

$$\nabla_a \nabla_b f - \nabla_b \nabla_a f = 0, \tag{1.108}$$

for all smooth functions f. It contains torsion if $\nabla_a \nabla_b f \neq \nabla_b \nabla_a f$. The torsion tensor is defined by

$$\nabla_a \nabla_b f - \nabla_b \nabla_a f = 2T_{ba}{}^i \partial_i f. \tag{1.109}$$

Our next task is to construct this covariant derivative explicitly. We know how it acts on scalars and we also know that in flat space it must correspond to partial differentiation. Let us then consider the object $\nabla_i A^n$. Recall that partial derivatives of vectors do not transform like tensors, so $\nabla_i A^n$ and $\partial_i A^n$ should differ by some quantity which also does not transform like a tensor under coordinate transformations. This difference should only depend on A^n and the geometry

---

[3]Literally translated from Latin it means 'from the one before' or 'from the earlier'.

of the space, hence we wish to write

$$\nabla_i A^n = \partial_i A^n + C_{ij}^n A^j \tag{1.110}$$

and need to determine the coefficients C_{ij}^n. These C_{ij}^n are called the connection coefficients or simply the connection.

First, we note that $\nabla_i(A^n B_n) = \partial_i(A^n B_n)$ since $A^n B_n$ is a scalar. Using the product rule $(\nabla_i A^n) B_n + A^n(\nabla_i B_n) = (\partial_i A^n) B_n + A^n(\partial_i B_n)$ we can find an expression for $\nabla_i B_n$ which is given by

$$\nabla_i B_n = \partial_i B_n - C_{in}^j B_j. \tag{1.111}$$

Furthermore, again using the product rule, we can deduce that the covariant derivative acts on each and every index separately.

One of the main results of Riemannian Geometry in the context of General Relativity is summarised in the following theorem, the proof of which is important to understand and based on a simple but powerful idea.

Theorem 1.1 (Uniqueness of the covariant derivative). *Let ∇_a be the covariant derivative on some manifold $\mathcal{M}$. If the covariant derivative is metric compatible and torsion-free, then the connection coefficients are uniquely given by the Christoffel symbol, $C_{ij}^n = \Gamma_{ij}^n$.*

Proof. Let us begin with $\nabla_a \nabla_b f$ and $\nabla_b \nabla_a f$ which written out explicitly give

$$\nabla_a \nabla_b f = \partial_a \partial_b f - C_{ab}^j \partial_j f, \tag{1.112}$$

$$\nabla_b \nabla_a f = \partial_b \partial_a f - C_{ba}^j \partial_j f. \tag{1.113}$$

One of our assumptions is that the covariant derivative is torsion-free which means that we assume $\nabla_a \nabla_b f = \nabla_b \nabla_a f$. Since partial derivatives also commute, we must conclude that $C_{ab}^j = C_{ba}^j$ which means the coefficients are symmetric in the lower pair of indices. Next, we use that the covariant derivative is metric compatible, this means

$\nabla_a g_{bc} = 0$. We write out this identity three times with permuted indices

$$\nabla_a g_{bc} = \partial_a g_{bc} - C^d_{ab} g_{dc} - C^d_{ac} g_{bd} = 0, \qquad (1.114)$$

$$\nabla_b g_{ca} = \partial_b g_{ca} - C^d_{bc} g_{da} - C^d_{ba} g_{cd} = 0, \qquad (1.115)$$

$$\nabla_c g_{ab} = \partial_c g_{ab} - C^d_{ca} g_{db} - C^d_{cb} g_{ad} = 0. \qquad (1.116)$$

The idea of the next step is to solve those three equations for the unknown coefficients. Let us now calculate (1.114) + (1.115) − (1.116) which gives

$$\partial_a g_{bc} + \partial_b g_{ca} - \partial_c g_{ab} - 2C^d_{ab} g_{dc} = 0, \qquad (1.117)$$

where we took into account that the metric is symmetric and that the coefficients are symmetric. We apply g^{nc} to this equation and isolate the connection. This gives

$$C^n_{ab} = \frac{1}{2} g^{nc} \left(\partial_a g_{bc} + \partial_b g_{ca} - \partial_c g_{ab} \right), \qquad (1.118)$$

which is indeed the desired result as the right-hand side matches Eq. (1.71) with different index names. $\qquad \square$

Therefore, we have a unique covariant derivative and its connection coefficients are given by the Christoffel symbol. It is useful to collect some important formulae when the covariant derivative acts on rank-1 and rank-2 tensors, these will be required regularly

$$\nabla_a A^b = \partial_a A^b + \Gamma^b_{ac} A^c, \qquad (1.119)$$

$$\nabla_a A_b = \partial_a A_b - \Gamma^c_{ab} A_c, \qquad (1.120)$$

$$\nabla_a T^{ij} = \partial_a T^{ij} + \Gamma^i_{ac} T^{cj} + \Gamma^j_{ac} T^{ic}, \qquad (1.121)$$

$$\nabla_a T^i_j = \partial_a T^i_j + \Gamma^i_{ac} T^c_j - \Gamma^c_{aj} T^i_c, \qquad (1.122)$$

$$\nabla_a T_{ij} = \partial_a T_{ij} - \Gamma^c_{ai} T_{cj} - \Gamma^c_{aj} T_{ic}. \qquad (1.123)$$

One can easily state the general formula for computing the covariant derivative of a rank-n tensor, however, we will not need this for what follows and hence omit this cumbersome equation. Before proceeding, we work through an example.

Example 1.12 (2D Laplacian in polar coordinates). Let us compute the quantity $g^{ij}\nabla_i\nabla_j f$ where f is a smooth function. We have

$$g^{ij}\nabla_i\nabla_j f = g^{ij}\nabla_i\partial_j f = g^{ij}\partial_i\partial_j f - g^{ij}\Gamma^n_{ij}\partial_n f. \qquad (1.124)$$

When working with Cartesian coordinates $(X^1, X^2) = (x, y)$, all Christoffel symbol components are zero and we would find

$$g^{ij}\nabla_i\nabla_j f = \partial_{xx}f + \partial_{yy}f = \Delta f. \qquad (1.125)$$

In polar coordinates $(X^1, X^2) = (r, \varphi)$, on the other hand, we have

$$
\begin{aligned}
g^{ij}\nabla_i\nabla_j f &= \partial_{rr}f + \frac{1}{r^2}\partial_{\varphi\varphi}f - \frac{1}{r^2}(-r)\partial_r f \\
&= \partial_{rr}f + \frac{1}{r}\partial_r f + \frac{1}{r^2}\partial_{\varphi\varphi}f, \qquad (1.126)
\end{aligned}
$$

which is the Laplacian Δf in polar coordinates.

1.2.7. *Parallel transport and geodesics*

Definition 1.18 (Parallel transport). Let C be a curve with tangent vector T^a, and let V^a be a vector. We say that V^a is parallelly transported along the curve if

$$T^a\nabla_a V^b = 0, \qquad (1.127)$$

at every point on the curve. The vector V^a can in principle be replaced by an arbitrary rank tensor.

The idea behind this definition is to transport the vector V^a along a curve and changing its orientation as little as possible. Recalling the chain rule $T^a\partial_a V^b = \dot{X}^a\partial_a V^b = dV^b/d\lambda$, we can alternatively write Eq. (1.127) in the form

$$\frac{dV^b}{d\lambda} = -T^a\Gamma^b_{ac}V^c, \qquad (1.128)$$

which is a system of first-order ordinary differential equations in the vector components V^i.

Example 1.13 (Parallel transport). Let us consider 2D Cartesian space with polar coordinates (r, φ) and the curve $X^a = (\lambda, k)$ with k being constant. The tangent vector to this curve is given by $T^a = (1, 0)$, so that a vector V^i parallelly transported along this curve has to satisfy the two equations

$$\dot{V}^1 = -T^a \Gamma^1_{ac} V^c = 0, \tag{1.129}$$

$$\dot{V}^2 = -\Gamma^2_{12} V^2 = -\frac{1}{r} V^2 = -\frac{1}{\lambda} V^2. \tag{1.130}$$

Therefore, $V^1 = \alpha$ and $V^2 = \beta/\lambda$ where α and β are constants which are fixed by the initial position of the vector V^i. Since $r(\lambda) = \lambda$, we can also write $V^i = (\alpha, \beta/r)$ and check directly whether Eq. (1.127) is indeed satisfied. We have

$$T^a \nabla_a V^1 = T^a \partial_a V^1 + T^a \Gamma^1_{ac} V^c = 0,$$

$$T^a \nabla_a V^2 = T^a \partial_a V^2 + T^a \Gamma^2_{ac} V^c = \frac{\partial}{\partial r} \frac{\beta}{r} + \frac{1}{r} \frac{\beta}{r} = 0. \tag{1.131}$$

Hence, V^i is indeed the vector parallelly transported along the curve.

Instead of transporting an arbitrary vector along a curve, we can also transport the tangent vector itself along the curve which defines it. Intuitively speaking this corresponds to the straightest possible lines, sometimes called autoparallels, which are not necessarily identical to our shortest possible lines, the geodesics mentioned in Sec. 1.2.5. It is an interesting fact that in general shortest lines and straightest lines differ. However, in spaces where the connection is metric and torsion-free they are the same.

Theorem 1.2. *Let ∇_a be a metric and torsion-free covariant deriva-tive, then a curve with tangent vector T^a satisfying the equation*

$$T^a \nabla_a T^b = 0, \tag{1.132}$$

is a geodesic.

Proof. We assume that our curve is parameterised by an affine parameter λ so that $X^i = X^i(\lambda)$ and $T^i = \dot{X}^i$. This gives

$$\dot{X}^a \partial_a \dot{X}^b + \dot{X}^a \Gamma^b_{ac} \dot{X}^c = 0. \tag{1.133}$$

The chain rule allows us to write

$$\frac{dX^a}{d\lambda} \frac{\partial}{\partial X^a} \frac{dX^b}{d\lambda} = \frac{d^2 X^b}{d\lambda^2} = \ddot{X}^b, \tag{1.134}$$

and we arrive at

$$\ddot{X}^b + \Gamma^b_{ac} \dot{X}^a \dot{X}^c = 0, \tag{1.135}$$

which is identical to the geodesic equation (1.70). □

Sometimes this fact is used to define geodesics using the concept of parallel transport.

Example 1.14 (Geodesics on the 2-sphere). The line element of the 2-sphere was given by $ds^2 = R^2(d\theta^2 + \sin^2 \theta d\phi^2)$, we use spherical polar coordinates $X^i = (\theta, \phi)$. We note that the azimuth $\phi \in [0, 2\pi)$ corresponds to the longitude. The geographic longitude is usually in the interval $[-\pi, \pi)$ and the sign denote the direction, West or East. Likewise the inclination is $\theta \in [0, \pi)$, and $[\pi/2, \pi/2)$ for the geographic latitude.

The Lagrangian is given by $L = R^2(\dot{\theta}^2 + \sin^2 \theta \dot{\phi}^2)$ so that we arrive at the following equations

$$\frac{\partial L}{\partial \theta} = \frac{d}{d\lambda} \frac{\partial L}{\partial \dot{\theta}} \quad \Rightarrow \quad R^2 \sin \theta \cos \theta \dot{\phi}^2 = \ddot{\theta}, \tag{1.136}$$

$$\frac{\partial L}{\partial \phi} = \frac{d}{d\lambda}\frac{\partial L}{\partial \dot{\phi}} \quad \Rightarrow \quad 0 = \frac{d}{d\lambda}\left(\sin^2\theta\dot{\phi}\right), \qquad (1.137)$$

which we rewrite slightly so that they take this form

$$\ddot{\theta} - R^2 \sin\theta\cos\theta\dot{\phi}^2 = 0, \qquad (1.138)$$

$$\ddot{\phi} + 2\cot\theta\dot{\phi}\dot{\theta} = 0. \qquad (1.139)$$

One can now read off the Christoffel symbol components

$$\Gamma_{11}^1 = 0, \quad \Gamma_{12}^1 = 0, \quad \Gamma_{22}^1 = -R^2 \sin\theta\cos\theta \qquad (1.140)$$

and likewise for the ϕ-equation

$$\Gamma_{11}^2 = 0, \quad \Gamma_{12}^2 = \cot\theta, \quad \Gamma_{22}^2 = 0. \qquad (1.141)$$

We will not solve these equations but note that solving them is similar to Example 1.10. Equation (1.137) yields conservation of angular momentum and can be used to eliminate $\dot{\phi}$ from Eq. (1.136). One arrives at great circles.

Since shortest and straightest lines are not necessarily the same, this also allows us to introduce the concept of the connection fully independent of the metric via parallel transport. One could state that the vector V^i changes according to $dV^b = -\Gamma_{ac}^b V^c dX^a$ when parallelly displaced from X^a to $X^a + dX^a$. A connection introduced in this way is often called an affine connection. In n dimensions it has n^3 independent components. On the other hand, we can use the metric to define the shortest distance between two points which gives us geodesics.

Definition 1.19 (Geodesics and autoparallels). Let $\mathcal{M}$ be a manifold with metric g_{ij} and independent connection Γ_{ij}^n. A curve X^b is called a geodesic if it satisfies

$$\ddot{X}^b + \left\{\begin{smallmatrix} b \\ ac \end{smallmatrix}\right\}\dot{X}^a\dot{X}^c = 0. \qquad (1.142)$$

A curve Y^b is called an autoparallel if it satisfies

$$\ddot{Y}^b + \Gamma_{ac}^b\dot{Y}^a\dot{Y}^c = 0. \qquad (1.143)$$

Exercise 1.14 shows that the connection is a space with torsion, is given by

$$\Gamma^k_{ij} = \{^k_{ij}\} + T_{ij}{}^k - T_j{}^k{}_i + T^k{}_{ij}. \tag{1.144}$$

Since $\dot{Y}^a\dot{Y}^c$ is symmetric, $\Gamma^b_{ac}\dot{Y}^a\dot{Y}^c$ depends only on the symmetric part of the connection. It is tempting to think that torsion will therefore not affect Eq. (1.143), however, this is not correct. Let us compute the symmetric part of the connection

$$\begin{aligned}
\Gamma^k_{(ij)} &= \frac{1}{2}\left(\Gamma^k_{ij} + \Gamma^k_{ji}\right) \\
&= \{^k_{ij}\} + \frac{1}{2}\left(T_{ij}{}^k - T_j{}^k{}_i + T^k{}_{ij} + T_{ji}{}^k - T_i{}^k{}_j + T^k{}_{ji}\right) \\
&= \{^k_{ij}\} + \left(T^k{}_{ij} + T^k{}_{ji}\right).
\end{aligned} \tag{1.145}$$

At first this is counter-intuitive as the skew-symmetric part of the connection is the torsion tensor, however, the symmetric part of the connection also depends on the torsion tensor and hence in spaces with torsion geodesics and autoparallels are different. In other words, shortest lines are different to straightest lines.

1.3. Curvature

We are now within touching distance of curvature. Let us consider a small (infinitesimal) parallelogram defined by two vectors at some point p. Now, we take a third vector and parallelly transport this vector around this parallelogram. In a flat space, transporting the vector around the parallelogram will not change the vector, however, in a curved space the change of this vector is related to a rank-4 tensor, the so-called Riemann curvature tensor, which is related to the Christoffel symbols. This section is entirely about this object.

Let us start with an example, the surface of a sphere, or for simplicity imagine the Earth. Start at the equator where it intersects the Greenwich Meridian (0° longitude) and, looking North, walk to the North pole, pretend to walk over the water. Keep your head fixed. At the North pole, walk back to the equator (walk sideways as we want to keep our orientation fixed) along the W90° meridian. Then walk back (this time backwards) along the equator to where we started. We should now be looking westwards, so we somehow managed to turn our body by 90° despite keeping our relative orientation fixed.

If we were to use some tape and a ball to trace our imaginary route out, we would also notice that we walked along a triangle with three right angles, this means 270°. This sounds like a contradiction, however, it all fits together very well. In Euclidean space the sum of the three angles in a triangle is 180°, in spaces with negative curvature this sum is less than 180°, and in spaces with positive curvature the sum of the angles is more than 180°. The surface of a sphere is a space of constant positive curvature.

1.3.1. *Infinitesimal parallelogram*

We begin with a vector V^a at a point p and coordinates X^i. We denote by ξ^i and ζ^i two infinitesimal quantities which we can view as the two directions which define a parallelogram, starting from the point $p(X^i)$. We also denote $p' = p(X^i + \xi^i)$ and $p'' = p(X^i + \zeta^i)$ and finally $q = p(X^i + \xi^i + \zeta^i)$, see Fig. 1.3. The following calculation is quite tricky and it is easy to get lost, so it is advisable to spend some time on this.

The equation of parallel transport of V^a along ξ^i can be written as

$$\xi^a \partial_a V^b = -\xi^a \Gamma^b_{ac} V^c, \qquad (1.146)$$

where V^b is the vector at p. However, for $V^b(X^i + \xi^i)$ we also have

$$V^b\big|_{p'} = V^b(X^i) + \xi^j \partial_j V^b(X^i) = V^b\big|_p + \xi^j \partial_j V^b\big|_p, \qquad (1.147)$$

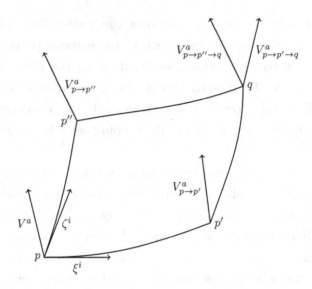

Fig. 1.3   Vector V^a parallelly transported along two different paths in an infinitesimal parallelogram.

where we used a Taylor series expansion and neglected terms of higher than linear order because we assumed ξ^i to be infinitesimal. Combining both equations we can now compute the difference of the vector at the two points p and p'. We define $\delta V^b = V^b\big|_{p'} - V^b\big|_p$ and find

$$\delta V^b = -\xi^a \Gamma^b_{ac}\big|_p V^c\big|_p. \tag{1.148}$$

Next, the new vector Eq. (1.147) is parallelly transported along the other way ζ^i towards q. This gives

$$\zeta^a \nabla_a V^b\big|_{p'} = \zeta^a \partial_a V^b\big|_{p'} + \zeta^a \Gamma^b_{ac}\big|_{p'} V^c\big|_{p'} = 0. \tag{1.149}$$

We have to be careful here as the Christoffel symbol is evaluated at the same point as the vector, this means at p'. Using again the Taylor series, we have

$$\zeta^a \partial_a V^b\big|_{p'} = \zeta^a \partial_a V^b\big|_p + \zeta^a \xi^c \partial_a \partial_c V^b\big|_p$$
$$= -\zeta^a \Gamma^b_{ac}\big|_p V^c\big|_p + \zeta^a \xi^c \partial_a \partial_c V^b\big|_p, \tag{1.150}$$

where we used Eq. (1.146) to re-write the first partial derivative term. We solve for the term with the second derivatives and substitute into Eq. (1.149). This yields

$$\zeta^a \xi^c \partial_a \partial_c V^b \big|_p = \zeta^a \Gamma^b_{ac}\big|_p V^c\big|_p - \zeta^a \Gamma^b_{ac}\big|_{p'} V^c\big|_{p'}. \tag{1.151}$$

We now perform another Taylor series expansion on the terms evaluated at p'. We note that the lowest-order terms cancel on the right-hand side and we arrive at

$$\begin{aligned}
\zeta^a \xi^c \partial_a \partial_c V^b \big|_p &= -\zeta^a \xi^d \Gamma^b_{ac}\big|_p \partial_d V^c\big|_p - \zeta^a \xi^d \partial_d \Gamma^b_{ac}\big|_p V^c\big|_p \\
&= \zeta^a \xi^d \Gamma^b_{ac}\big|_p \Gamma^c_{df} V^f\big|_p - \zeta^a \xi^d \partial_d \Gamma^b_{af}\big|_p V^f\big|_p.
\end{aligned} \tag{1.152}$$

As everything is now evaluated at the same point p, we can drop this label and simply write

$$\zeta^a \xi^c \partial_a \partial_c V^b = \zeta^a \xi^d \left(\Gamma^b_{ac} \Gamma^c_{df} - \partial_d \Gamma^b_{af} \right) V^f. \tag{1.153}$$

We could have parallelly transported along ζ^i first and secondly along the ξ^i direction. This would have resulted in Eq. (1.153) with ξ^i and ζ^i interchanged. We are interested in the difference between those two vectors parallelly transported along the two parts of the parallelogram

$$\Delta V^b = V^b_{p \to p'' \to q} - V^b_{p \to p' \to q}. \tag{1.154}$$

This quantity is given by

$$\Delta V^b = \zeta^a \xi^d \left(\partial_d \Gamma^b_{af} - \partial_a \Gamma^b_{df} + \Gamma^b_{dc} \Gamma^c_{af} - \Gamma^b_{ac} \Gamma^c_{df} \right) V^f. \tag{1.155}$$

The term in the bracket is the so-called Riemann curvature tensor, a rank-4 tensor. However, before discussing this tensor in detail in the next section, let us make a few observations about Eq. (1.155).

The change of the vector V^b when parallelly transported around the parallelogram depends on the derivatives of the Christoffel symbols and terms quadratic in the Christoffel symbols, there are no linear terms. It is not clear at this point that the term in the bracket is really a tensor, it does contain combinations of partial derivatives and Christoffel symbols similar to those in the covariant derivative, however, this needs to be checked. Moreover, there are some obvious symmetries, interchanging the indices d and a will change the sign of the expression in the bracket.

This is also a good point to ask the following question: Begin with two vectors ξ^i and ζ^i and parallelly transport ξ^i along ζ^i and then ζ^i along ξ^i. Does this process result in a parallelogram? The answer is a somewhat surprising no, in general. Starting with Eq. (1.148) we would find

$$\delta\xi^b = -\zeta^a \Gamma^b_{ac} \xi^c, \qquad (1.156)$$

$$\delta\zeta^b = -\xi^a \Gamma^b_{ac} \zeta^c. \qquad (1.157)$$

Hence the difference between those two is

$$\delta\xi^b - \delta\zeta^b = \xi^i \zeta^j (\Gamma^b_{ij} - \Gamma^b_{ji}). \qquad (1.158)$$

The right-hand side is zero if and only if the connection is symmetric in the pair of lower indices. This relates to Definition 1.17 of the torsion tensor and the interpretation that the torsion tensor measures the failure of a parallelogram to close. For a given general connection Γ^b_{ij} the torsion tensor is defined to be the skew-symmetric part of the connection

$$T_{ij}{}^k = \frac{1}{2}\left(\Gamma^k_{ij} - \Gamma^k_{ji}\right). \qquad (1.159)$$

This relation follows from Eq. (1.109) by writing out the covariant derivatives using the Christoffel symbol. Despite the connection not transforming like a tensor, we should note that torsion is always a tensor since the inhomogeneous parts in the transformation will cancel, similar to Eq. (1.39).

1.3.2. *Riemann, Ricci and Weyl tensors*

We already mentioned that the term in the bracket of Eq. (1.155) looks like a covariant derivative acting on a Christoffel symbols. We will now derive this term differently. Let us start by computing $\nabla_a \nabla_d V^b$ and keeping in mind that $\nabla_d V^b$ is a rank-2 tensor. We have

$$
\begin{aligned}
\nabla_a \nabla_d V^b &= \partial_a (\nabla_d V^b) - \Gamma^i_{ad}(\nabla_i V^b) + \Gamma^b_{aj}(\nabla_d V^j) \\
&= \partial_a \partial_d V^b + \partial_a \Gamma^b_{di} V^i + \Gamma^b_{di} \partial_a V^i - \Gamma^i_{ad} \partial_i V^b \\
&\quad - \Gamma^i_{ad} \Gamma^b_{ij} V^j + \Gamma^b_{aj} \partial_d V^j - \Gamma^b_{aj} \Gamma^j_{di} V^i.
\end{aligned}
\tag{1.160}
$$

Next, we write out $\nabla_d \nabla_a V^b$ and subtract both terms from each other, in other words we are computing the commutator of two covariant derivatives. It becomes clear that quite a few terms will cancel during this calculation, the result of which is

$$
(\nabla_a \nabla_d - \nabla_d \nabla_a) V^b = \left(\partial_a \Gamma^b_{di} - \partial_d \Gamma^b_{ai} + \Gamma^b_{dj} \Gamma^j_{ai} - \Gamma^b_{aj} \Gamma^j_{di} \right) V^i.
\tag{1.161}
$$

Comparing the right-hand side of Eq. (1.161) with Eq. (1.155) we note that these term only differ by a minus sign and some renaming of dummy indices. The good news is that Eq. (1.161) is clearly a tensor-valued object because we defined the covariant derivative precisely in this way, it maps tensors to tensors. This in turn implies that the parallel transport of a vector along a small parallelogram is determined by a tensorial quantity. We can promote this into a definition.

Definition 1.20 (Riemann curvature tensor). Let ∇_a be a covariant derivative and let V^b be a vector. The equation

$$
\nabla_a \nabla_d V^b - \nabla_d \nabla_a V^b = -R_{adi}{}^b V^i,
\tag{1.162}
$$

defines the Riemann curvature tensor (short Riemann tensor) $R_{adi}{}^b$.

This means we can write Eq. (1.155) as

$$\Delta V^b = \zeta^a \xi^d R_{adf}{}^b V^f, \qquad (1.163)$$

and for completeness we write out the Riemann tensor explicitly

$$R_{adf}{}^b = \partial_d \Gamma^b_{af} - \partial_a \Gamma^b_{df} + \Gamma^b_{dc}\Gamma^c_{af} - \Gamma^b_{ac}\Gamma^c_{df}. \qquad (1.164)$$

We note that there are different conventions when defining the Riemann curvature tensor, for instance, the minus sign in Eq. (1.162). This sign depends on how one defines the tensor. Had we started with a covariant vector V_b, we would have found that

$$\nabla_a \nabla_d V_b - \nabla_d \nabla_a V_b = R_{adb}{}^i V_i. \qquad (1.165)$$

This sign is related to the different signs when applying the covariant derivative to upper or lower indices. Analogously to the covariant derivative one can generalise formulae (1.164) and (1.165) to higher-rank tensors.

Let us briefly revisit Eq. (1.39). We defined $F_{ij} = \nabla_i A_j - \nabla_j A_i$, it turns out that the Christoffel symbols coming from the covariant derivatives will cancel. A direct calculation shows

$$\nabla_i A_j - \nabla_j A_i = \partial_i A_j - \Gamma^n_{ij} A_n - \left(\partial_j A_i - \Gamma^n_{ji} A_n \right)$$
$$= \partial_i A_j - \partial_j A_i, \qquad (1.166)$$

which leads us to a nice little result which will be needed.

Theorem 1.3. *For $F_{ij} = \nabla_i A_j - \nabla_j A_i = -F_{ji}$ the following identity holds*

$$\nabla_i F_{jk} + \nabla_k F_{ij} + \nabla_j F_{ki} = 0. \qquad (1.167)$$

Proof. We begin with

$$\nabla_i F_{jk} = \partial_i F_{jk} - \Gamma^n_{ij} F_{nk} - \Gamma^n_{ik} F_{jn}, \qquad (1.168)$$

$$\nabla_k F_{ij} = \partial_k F_{ij} - \Gamma^n_{ki} F_{nj} - \Gamma^n_{kj} F_{in}, \tag{1.169}$$

$$\nabla_j F_{ki} = \partial_j F_{ki} - \Gamma^n_{jk} F_{ni} - \Gamma^n_{ji} F_{kn}. \tag{1.170}$$

We add up these three equations and use that $F_{ij} = -F_{ji}$ and $\Gamma^n_{ij} = \Gamma^n_{ji}$ which gives

$$\nabla_i F_{jk} + \nabla_k F_{ij} + \nabla_j F_{ki} = \partial_i F_{jk} + \partial_k F_{ij} + \partial_j F_{ki}, \tag{1.171}$$

again the Christoffel symbols all cancel. Since $F_{ij} = \partial_i A_j - \partial_j A_i$ and partial derivatives commute, we arrive at the identity. $\qquad\square$

The Riemann tensor satisfies some important identities which are summarised in a theorem. The proofs are based on direct calculations and are a good exercise to get used to working with the Riemann tensor. The basic idea is always the same, namely writing out the definition of the Riemann curvature tensor with permuted indices.

Theorem 1.4. *Let R_{abcd} be the Riemann curvature tensor defined by a torsion-free and metric compatible covariant derivative. R_{abcd} satisfies the following identities:*

(i) $R_{abcd} = -R_{bacd}$.

(ii) $R_{abcd} = -R_{abdc}$.

(iii) $R_{abcd} + R_{cabd} + R_{bcad} = 0$.

(iv) $\nabla_a R_{bcde} + \nabla_c R_{abde} + \nabla_b R_{cade} = 0$ *(Bianchi identity)*.

Proof. (i) This follows from the definition, see for instance Eq. (1.165).

(ii) We start by recalling that $\nabla_a g_{bc} = 0$. Now we compute $\nabla_a \nabla_b g_{cd} - \nabla_b \nabla_a g_{cd}$ and express this using the Riemann curvature tensor, we have

$$0 = \nabla_a \nabla_b g_{cd} - \nabla_b \nabla_a g_{cd} = R_{abc}{}^i g_{id} + R_{abd}{}^i g_{ci} = R_{abcd} + R_{abdc}, \tag{1.172}$$

which implies the second symmetry property.

(iii) We proceed similarly to the proof of Theorem 1.1. We write out $\nabla_a \nabla_b V_c - \nabla_b \nabla_a V_c$ three times with permuted indices

$$\nabla_a \nabla_b V_c - \nabla_b \nabla_a V_c = R_{abc}{}^i V_i, \qquad (1.173)$$

$$\nabla_c \nabla_a V_b - \nabla_a \nabla_c V_b = R_{cab}{}^i V_i, \qquad (1.174)$$

$$\nabla_b \nabla_c V_a - \nabla_c \nabla_b V_a = R_{bca}{}^i V_i. \qquad (1.175)$$

These three equations are added up next, and we can now use the previous theorem by introducing the notation $\mathcal{F}_{ab} = \nabla_a V_b - \nabla_b V_a$ and arrive at

$$\nabla_a \mathcal{F}_{bc} + \nabla_c \mathcal{F}_{ab} + \nabla_b \mathcal{F}_{ca} = \left(R_{abc}{}^i + R_{cab}{}^i + R_{bca}{}^i \right) V_i. \qquad (1.176)$$

We know that the left-hand side of Eq. (1.176) vanishes due to Theorem 1.3. Since this equation is valid for all vectors V_i, we have proved the third property.

(iv) Since $\nabla_i V_j$ is a rank-2 tensor, we have

$$\nabla_a \nabla_b \nabla_i V_j - \nabla_b \nabla_a \nabla_i V_j = R_{abi}{}^n (\nabla_n V_j) + R_{abj}{}^n (\nabla_i V_n). \qquad (1.177)$$

On the other hand, we have Eq. (1.165) and by applying ∇_a to this equation we arrive at the second required equation

$$\nabla_a \nabla_b \nabla_i V_j - \nabla_a \nabla_i \nabla_b V_j = (\nabla_a R_{bij}{}^n) V_n + R_{bij}{}^n \nabla_a V_n, \qquad (1.178)$$

where we used the product rule on the right-hand side. The term $\nabla_a R_{bij}{}^n$ is the object we need for the Bianchi identity. Hence, we want to write out Eqs. (1.177) and (1.178) three times with permuted indices. Starting with Eq. (1.177) we have

$$\nabla_a \nabla_b \nabla_i V_j - \nabla_b \nabla_a \nabla_i V_j = R_{abi}{}^n (\nabla_n V_j) + R_{abj}{}^n (\nabla_i V_n), \qquad (1.179)$$

$$\nabla_i \nabla_a \nabla_b V_j - \nabla_a \nabla_i \nabla_b V_j = R_{iab}{}^n (\nabla_n V_j) + R_{iaj}{}^n (\nabla_b V_n), \qquad (1.180)$$

$$\nabla_b \nabla_i \nabla_a V_j - \nabla_i \nabla_b \nabla_a V_j = R_{bia}{}^n (\nabla_n V_j) + R_{bij}{}^n (\nabla_a V_n). \qquad (1.181)$$

Likewise, from Eq. (1.178) one gets

$$\nabla_a\nabla_b\nabla_iV_j - \nabla_a\nabla_i\nabla_bV_j = (\nabla_aR_{bij}{}^n)V_n + R_{bij}{}^n\nabla_aV_n, \quad (1.182)$$

$$\nabla_i\nabla_a\nabla_bV_j - \nabla_i\nabla_b\nabla_aV_j = (\nabla_iR_{abj}{}^n)V_n + R_{abj}{}^n\nabla_iV_n, \quad (1.183)$$

$$\nabla_b\nabla_i\nabla_aV_j - \nabla_b\nabla_a\nabla_iV_j = (\nabla_bR_{iaj}{}^n)V_n + R_{iaj}{}^n\nabla_bV_n. \quad (1.184)$$

Now, if we calculate (1.179) + (1.180) + (1.181) and separately (1.182) + (1.183) + (1.184) we note that the left-hand sides of these new equations coincide, hence also their right-hand sides must be equal. Due to property (iii) the terms involving $R_{abi}{}^n$ will cancel and we find the relation

$$\begin{aligned}R_{abj}{}^n(\nabla_iV_n) &+R_{iaj}{}^n(\nabla_bV_n) + R_{bij}{}^n(\nabla_aV_n) \\ &= (\nabla_aR_{bij}{}^n)V_n + (\nabla_iR_{abj}{}^n)V_n + (\nabla_bR_{iaj}{}^n)V_n \\ &\quad +R_{bij}{}^n\nabla_aV_n + R_{abj}{}^n\nabla_iV_n + R_{iaj}{}^n\nabla_bV_n. \quad (1.185)\end{aligned}$$

At this point we note that all terms containing covariant derivatives of the vector V_k cancel and we are left with

$$(\nabla_aR_{bij}{}^n + \nabla_iR_{abj}{}^n + \nabla_bR_{iaj}{}^n)V_n = 0, \quad (1.186)$$

which must hold for all vectors V_n and proves the Bianchi identity.

$\square$

Theorem 1.5. *Let W_{abcd} be a tensor satisfying* (i) $W_{abcd} = -W_{bacd}$, (ii) $W_{abcd} = -W_{abdc}$ *and* (iii) $W_{abcd} + W_{cabd} + W_{bcad} = 0$, *then W_{abcd} also satisfies*

$$W_{abcd} = W_{cdab}. \quad (1.187)$$

This means that the three properties (i)–(iii) *of the Riemann curvature tensor imply a fourth algebraic identity.*

Proof. The proof is set as Exercise 1.19 and the solution will be given explicitly. The idea of the proof is to write our identity (iii) four times with permuted indices and to isolate the required two terms which make up the fourth algebraic identity.   □

A rank-4 object with no symmetries has n^4 independent components in general. However, the Riemann curvature tensor has various symmetry properties and hence the number of independent components is not obvious. It turns out that in dimensions $2, 3, 4$, the Riemann tensor has $1, 6, 20$ components, respectively. The general formula is $n^2(n^2 - 1)/12$ and its derivation is given as Exercise 1.20.

Since the Riemann tensor is a rank-4 tensor, we can consider its contractions (traces) over pairs of indices. Since it is skew-symmetric in the first and last pair of indices, we cannot contract over those indices. However, we can contract over one index from the first pair and one from the second pair.

Definition 1.21 (Ricci tensor and Ricci scalar). The Ricci tensor is defined by

$$R_{ab} = R_{acb}{}^c, \qquad (1.188)$$

and is a symmetric tensor due to Theorem 1.5. In n dimensions this tensor has $n(n + 1)/2$ independent components.

The Ricci scalar or scalar curvature is defined by

$$R = R_c{}^c. \qquad (1.189)$$

In four dimensions, for instance, the Ricci tensor has 10 independent components while the Riemann tensor has 20. The corresponding trace-free part of the Riemann curvature tensor is called the Weyl tensor, or the conformal tensor. We mention its definition for sake of completeness but will not discuss the Weyl tensor in detail henceforth.

Definition 1.22 (Weyl or conformal tensor). The Weyl tensor is

$$C_{abcd} = R_{abcd} - \frac{2}{n-2}\left(g_{a[c}R_{d]b} - g_{b[c}R_{d]a}\right)$$
$$+ \frac{2}{(n-2)(n-1)}Rg_{a[c}g_{d]b}, \tag{1.190}$$

where we recall the notation introduced in Eq. (1.37).

Example 1.15 (Hyperbolic disk). We write the line element of Poincaré's hyperbolic disk with coordinates $X^i = \{\chi, \varphi\}$ in the form

$$ds^2 = d\chi^2 + \sinh^2\chi d\varphi^2, \tag{1.191}$$

and want to compute the Riemann tensor, Ricci tensor and the Ricci scalar. In two dimensions the Riemann tensor has one independent component, so it suffices to compute R_{1212}. We begin with computing the Christoffel symbol components using the Lagrangian approach

$$L = \dot{\chi}^2 + \sinh^2\chi\dot{\varphi}^2. \tag{1.192}$$

The geodesic equations are given by

$$\ddot{\chi} - \sinh\chi\cosh\chi\dot{\varphi}^2 = 0, \tag{1.193}$$
$$\ddot{\varphi} + 2\coth\chi\dot{\chi}\dot{\varphi} = 0 \tag{1.194}$$

and hence the Christoffel symbol components from the $\ddot{\chi}$ equations are

$$\Gamma^1_{11} = 0, \quad \Gamma^1_{12} = 0, \quad \Gamma^1_{22} = -\sinh\chi\cosh\chi, \tag{1.195}$$

while the $\ddot{\varphi}$ yields

$$\Gamma^2_{11} = 0, \quad \Gamma^2_{12} = \coth\chi, \quad \Gamma^2_{22} = 0. \tag{1.196}$$

The formula for $R_{121}{}^2$ is

$$R_{121}{}^2 = \partial_2\Gamma^2_{11} - \partial_1\Gamma^2_{21} + \Gamma^2_{2c}\Gamma^c_{11} - \Gamma^2_{1c}\Gamma^c_{21}$$

$$= -\frac{\partial}{\partial\chi}\coth\chi - (\coth\chi)^2$$

$$= \frac{1}{\sinh^2\chi} - \frac{\cosh^2\chi}{\sinh^2\chi} = -1. \tag{1.197}$$

By contracting over the second and fourth index, we arrive at the Ricci tensor with components $R_{11} = -1$, $R_{22} = \sinh^2\chi$ and zero otherwise. This means in particular that the Ricci tensor is proportional to the metric tensor, $R_{ij} = -g_{ij}$. Moreover, the Ricci scalar is simply $R = -2$. This justifies calling the hyperbolic disk a space of constant negative curvature.

One tensor of great importance for General Relativity is the Einstein tensor G_{ab} which forms the basis of the Einstein field equations.

Definition 1.23 (Einstein tensor). The Einstein tensor is given by

$$G_{ab} = R_{ab} - \frac{1}{2}R\,g_{ab}, \tag{1.198}$$

and is a symmetric rank-2 tensor. In n dimensions this tensor has $n(n+1)/2$ independent components, so in four dimensions it has 10 independent components.

Theorem 1.6. *The Einstein tensor satisfies*

$$\nabla_a G^a{}_b = 0. \tag{1.199}$$

Proof. We start with the Bianchi identities

$$\nabla_a R_{bcd}{}^e + \nabla_c R_{abd}{}^e + \nabla_b R_{cad}{}^e = 0. \tag{1.200}$$

First, we contract over the indices a and e which gives

$$\nabla_a R_{bcd}{}^a - \nabla_c R_{bd} + \nabla_b R_{cd} = 0, \tag{1.201}$$

where we used $R_{abd}{}^a = -R_{bad}{}^a = -R_{bd}$. Next, we apply g^{cd} and get

$$-\nabla_a R_b{}^a - \nabla_c R_b{}^c + \nabla_b R = 0. \tag{1.202}$$

We relabel one index, take into account that $\nabla_b = \nabla_c \delta_b^c$ and arrive at

$$-\nabla_c \left(R_b{}^c + R_b{}^c - \delta_b^c R \right) = 0,$$
$$\Leftrightarrow \quad -2\nabla_c \left(R_b{}^c - \frac{1}{2}\delta_b^c R \right) = 0, \tag{1.203}$$

which implies the property $\nabla_c G_b{}^c = 0$. One refers to this argument as the twice contracted Bianchi identities imply this property. □

1.3.3. *Geodesic deviation equation*

In the following we derive the so-called geodesic deviation equation. The idea is to understand the behaviour of nearby geodesics, and we expect the curvature of the manifold to determine this. Let us start with a surface $X^a(\lambda, s)$, we assume that for a fixed value $s = s_0$, the curves $X^a(\lambda, s_0)$ are geodesics with affine parameter λ. The tangent vectors to the geodesics are given by

$$T^a = \frac{dX^a}{d\lambda}, \tag{1.204}$$

and the vectors connecting nearby geodesics are given by

$$N^a = \frac{dX^a}{ds}. \tag{1.205}$$

We can always choose λ and s such that $g_{ab}T^a N^b = 0$.

From Eqs. (1.204) and (1.205) we get

$$\frac{d^2 X^a}{d\lambda ds} = \frac{dT^a}{ds} = \frac{\partial T^a}{\partial X^i}\frac{\partial X^i}{\partial s} = N^i \frac{\partial T^a}{\partial X^i}, \tag{1.206}$$

$$\frac{d^2 X^a}{ds d\lambda} = \frac{dN^a}{d\lambda} = \frac{\partial N^a}{\partial X^i}\frac{\partial X^i}{\partial \lambda} = T^i \frac{\partial N^a}{\partial X^i}. \tag{1.207}$$

Since partial derivatives commute these last two expressions must be equal. Furthermore, the Christoffel symbols are symmetric in the lower pair of indices which then implies

$$T^i \nabla_i N^a = N^i \nabla_i T^a. \tag{1.208}$$

If we view N^a as the vector connecting nearby geodesics, then $T^i \nabla_i N^a$ would describe its rate of change along the geodesic with tangent vector T^i. Hence we can interpret $T^i \nabla_i N^a$ as the relative velocity between nearby geodesics. Therefore, the object $T^j \nabla_j (T^i \nabla_i N^a)$ would correspond to the relative acceleration which we are going to compute.

$$\begin{aligned} T^j \nabla_j (T^i \nabla_i N^a) &= T^j \nabla_j (N^i \nabla_i T^a) \\ &= T^j \nabla_j N^i \nabla_i T^a + T^j N^i \nabla_j \nabla_i T^a \\ &= T^j \nabla_j N^i \nabla_i T^a + T^j N^i \nabla_i \nabla_j T^a - T^j N^i R_{jic}{}^a T^c, \end{aligned} \tag{1.209}$$

where in the second step we used Eq. (1.162).

Next, we rewrite the first two terms as follows:

$$\begin{aligned} T^j \nabla_j N^i \nabla_i T^a &+ T^j N^i \nabla_i \nabla_j T^a \\ &= N^j \nabla_j T^i \nabla_i T^a + T^j N^i \nabla_i \nabla_j T^a \\ &= N^j \nabla_j T^i \nabla_i T^a + N^j T^i \nabla_j \nabla_i T^a \\ &= N^j \nabla_j \left(T^i \nabla_i T^a \right) = 0, \end{aligned} \tag{1.210}$$

because we assumed that $X^a(\lambda, s_0)$ are geodesics which means $T^i \nabla_i T^a = 0$ by Theorem 1.2. Consequently we find

$$T^j \nabla_j (T^i \nabla_i N^a) = -R_{jic}{}^a T^j N^i T^c = R_{ijc}{}^a T^j T^c N^i, \tag{1.211}$$

which is the geodesic deviation equation. Sometimes one introduces the notation

$$\frac{DN^a}{D\lambda} = T^i \nabla_i N^a, \tag{1.212}$$

this is the derivative along the geodesic so that the geodesic equation can be written as

$$\frac{D^2 N^a}{D\lambda^2} = (R_{ijc}{}^a T^j T^c) N^i, \qquad (1.213)$$

and we could define the relative velocity by $V^i = DN^i/D\lambda$ and the relative acceleration by $A^i = DV^i/D\lambda$, in analogy with mechanics.

The geodesic deviation equation contains the second clue towards a geometrical theory of gravity which we will discuss in Sec. 2.2.2.

1.4.   Euler–Lagrange Equations

In classical mechanics the action is defined as

$$S = \int_{t_1}^{t_2} L(\dot{q}, q, t)dt, \qquad (1.214)$$

where L is called the Lagrangian which is a function of velocity $\dot{q}$, position q and time t. The Lagrangian has units of energy, and in classical mechanics it is the difference between kinetic energy and potential energy $L = T - V$. Hence, the action S has units of energy times time which happens to have the same dimension as angular momentum. The equations of motion of the system described by the Lagrangian L are found by using the calculus of variations, one wishes to find q such that S is a stationary point. In order to find such q, we begin by considering a small change in the position $q + \delta q$ with $\delta q \ll 1$, however, we will not allow changes to q at the end points, this means $\delta q(t_1) = \delta q(t_2) = 0$. Then, to first order in δq we find

$$L(\dot{q} + \dot{\delta}, q + \delta q, t) = L(\dot{q}, q, t) + \frac{\partial L}{\partial q}\delta q + \frac{\partial L}{\partial \dot{q}}\delta\dot{q} + \cdots. \qquad (1.215)$$

In the following, we will only keep terms up to first order in δq. Let us denote the change in S due to the change in the position by δS

so that

$$S + \delta S = \int_{t_1}^{t_2} \left[L(\dot{q}, q, t) + \frac{\partial L}{\partial q} \delta q + \frac{\partial L}{\partial \dot{q}} \delta \dot{q} \right] dt, \tag{1.216}$$

$$= \int_{t_1}^{t_2} L(\dot{q}, q, t) dt + \int_{t_1}^{t_2} \left[\frac{\partial L}{\partial q} \delta q + \frac{\partial L}{\partial \dot{q}} \delta \dot{q} \right] dt, \tag{1.217}$$

which means we can now write δS explicitly as

$$\delta S = \int_{t_1}^{t_2} \left[\frac{\partial L}{\partial q} \delta q + \frac{\partial L}{\partial \dot{q}} \delta \dot{q} \right] dt. \tag{1.218}$$

As we wish to write the integrand using the small quantity δq we will use integration by parts on the second term by writing

$$\frac{\partial L}{\partial \dot{q}} \delta \dot{q} = \frac{d}{dt} \left(\frac{\partial L}{\partial \dot{q}} \delta q \right) - \left(\frac{d}{dt} \frac{\partial L}{\partial \dot{q}} \right) \delta q. \tag{1.219}$$

The first term in this is a total derivative and will not contribute to the integral because we keep the end points fixed, therefore

$$\delta S = \int_{t_1}^{t_2} \left[\frac{\partial L}{\partial q} \delta q - \left(\frac{d}{dt} \frac{\partial L}{\partial \dot{q}} \right) \delta q \right] dt \tag{1.220}$$

$$= \int_{t_1}^{t_2} \left[\frac{\partial L}{\partial q} - \left(\frac{d}{dt} \frac{\partial L}{\partial \dot{q}} \right) \right] \delta q dt. \tag{1.221}$$

The action is stationary if $\delta S = 0$ which implies the Euler–Lagrange equations

$$\frac{\partial L}{\partial q} - \left(\frac{d}{dt} \frac{\partial L}{\partial \dot{q}} \right) = 0. \tag{1.222}$$

For a given Lagrangian L the Euler–Lagrange equations are the equations of motion of the physical system.

Example 1.16 (Harmonic oscillator). Let us have a quick look at the standard example of a harmonic oscillator with kinetic energy $T = m\dot{q}^2/2$ and potential energy $V = kq^2/2$ so that

$L = m\dot{q}^2/2 - kq^2/2$. We have

$$\frac{\partial L}{\partial q} = -kq, \quad \frac{\partial L}{\partial \dot{q}} = m\dot{q}, \quad \frac{d}{dt}\frac{\partial L}{\partial \dot{q}} = m\ddot{q}, \tag{1.223}$$

so that the Euler–Lagrange equation is given by

$$m\ddot{q} + kq = 0. \tag{1.224}$$

This is indeed the differential equation of the harmonic oscillator with spring constant k and particle mass m.

The Euler–Lagrange equation can easily be generalised to systems with more degrees of freedom. If we consider a system with i particles with positions q_i and velocities $\dot{q}_i$, respectively, then the system will be governed by i differential equations given by

$$\frac{\partial L}{\partial q_i} - \left(\frac{d}{dt}\frac{\partial L}{\partial \dot{q}_i} \right) = 0, \tag{1.225}$$

which means we have one Euler–Lagrange equation per particle.

When studying classical field theories like electromagnetism, we can no longer use point particles to describe them, but need to study fields. This means our variables will also become functions of the spatial coordinates. In classical mechanics all quantities are functions of time only. In the case of field theories we begin with a field $\psi = \psi(t, x, y, z)$. Its action is

$$S = \int_\Omega \mathcal{L}(\partial_i \psi, \psi, t) d^3x \, dt, \tag{1.226}$$

where $\partial_i \psi$ stands for all possible (first) partial derivatives of the field ψ. The volume is denoted by Ω. If we insist on S having dimensions of energy multiplied by time, then $\mathcal{L}d^3x$ must have units of energy and in turn $\mathcal{L}$ must have units of energy density, this means energy per volume. For this reason $\mathcal{L}$ is generally called the Lagrangian density while the Lagrangian would be

$$L = \int_V \mathcal{L}(\partial_i \psi, \psi, t) d^3 x, \tag{1.227}$$

which is a function of time only.

The derivation of the Euler–Lagrange equations is analogous to the single particle classical mechanics case and is based on considering a small change $\delta\psi$ which is kept fixed at the boundary $\partial\Omega$. The result is given by

$$\frac{\partial\mathcal{L}}{\partial\psi} - \frac{\partial}{\partial t}\frac{\partial\mathcal{L}}{\partial(\partial_t\psi)} - \frac{\partial}{\partial x}\frac{\partial\mathcal{L}}{\partial(\partial_x\psi)} - \frac{\partial}{\partial y}\frac{\partial\mathcal{L}}{\partial(\partial_y\psi)} - \frac{\partial}{\partial z}\frac{\partial\mathcal{L}}{\partial(\partial_z\psi)} = 0,$$
$$\tag{1.228}$$

and the presence of the extra terms simply follows from integration by parts with respect to the other variables. We can write this in a much neater way using the index notation and find

$$\frac{\partial\mathcal{L}}{\partial\psi} - \frac{\partial}{\partial x^i}\frac{\partial\mathcal{L}}{\partial(\partial_{x^i}\psi)} = 0. \tag{1.229}$$

Lastly, if the Lagrangian density depends on several different fields ψ_A, then we would have A Euler–Lagrange equations, one for each and every field similar to Eq. (1.225).

1.5.   Further Reading

One particular issue that was missed out is the so-called Lie derivative. This is a connection independent derivative operator. It provides a measure of the change of a vector (or tensor in general) along the flow of another vector field. Let U and V be two vectors, then the Lie derivative of V with respect to U is defined by $\mathcal{L}_U V^a = U^b \partial_b V^a - V^b \partial_b U^a$. This expression transforms like a tensor, similar to Eq. (1.39). The Lie derivative is a useful tool for more advanced topics of General Relativity and Cosmology. It is discussed in more detail in the recommended books.

Differential Geometry itself is a large research field in Mathematics, there are plenty of textbooks written which cover this field. Three well-known books aimed primarily at mathematicians are by Eisenhart (1997), a classic originally published in 1926, and the books by Bishop and Goldberg (1980) and do Carmo (1992) both of which make for an interesting and useful read. It turns out though that the index notation used for most parts of this book has been superseded in Mathematics by the use of differential forms and the so-called index-free notation.

However, most of the theoretical physics literature is written in the index notation. Other recommended books for further reading very much reflect this point, they are more aimed at theoretical physicists than mathematicians. The main theme in all of them is the interplay between Physics and Geometry. The books by Isham (2001) and Nakahara (2003) both aim to introduce the concepts of modern differential geometry to the reader while making links with physics. The most comprehensive book which discusses the concepts of Geometry and Physics in great detail is by Frankel (2012). Its almost 750 pages cover every important physical theory using the language of geometry. It is a great book but requires dedication from the reader.

The recommended literature is by no means complete, it simply reflects the author's suggestions for further reading.

1.6. Exercises

The concept of a vector

Exercise 1.1. Let a, b, c be given vectors which do not all lie in the same plane. Show that the volume of the parallelepiped spanned by these vectors is given by $V = |a \cdot (b \times c)|$.

Exercise 1.2. Show that the so-called scalar triple product satisfies $\boldsymbol{a} \cdot (\boldsymbol{b} \times \boldsymbol{c}) = \boldsymbol{c} \cdot (\boldsymbol{a} \times \boldsymbol{b}) = \boldsymbol{b} \cdot (\boldsymbol{c} \times \boldsymbol{a})$.

Exercise 1.3. What is $\boldsymbol{a} \cdot (\boldsymbol{b} \times \boldsymbol{c})$ in index notation?

Exercise 1.4. One can write $\varepsilon_{ijk}\varepsilon^{lmn}$ as follows:

$$\varepsilon_{ijk}\varepsilon^{lmn} = \det \begin{pmatrix} \delta_i^l & \delta_i^m & \delta_i^n \\ \delta_j^l & \delta_j^m & \delta_j^n \\ \delta_k^l & \delta_k^m & \delta_k^n \end{pmatrix}. \tag{1.230}$$

Here det stands for the determinant of the matrix. Use this to prove the following three identities:

$$\varepsilon_{abc}\varepsilon^{mnc} = \delta_a^m \delta_b^n - \delta_a^n \delta_b^m, \tag{1.231}$$

$$\varepsilon_{abc}\varepsilon^{mbc} = 2\delta_a^m, \tag{1.232}$$

$$\varepsilon_{abc}\varepsilon^{abc} = 6. \tag{1.233}$$

Exercise 1.5. Show that $\boldsymbol{a} \times (\boldsymbol{b} \times \boldsymbol{c}) = (\boldsymbol{a} \cdot \boldsymbol{c})\boldsymbol{b} - (\boldsymbol{a} \cdot \boldsymbol{b})\boldsymbol{c}$. Show this again by using the index notation.

Exercise 1.6. Show that $\varepsilon_{ijk} = (\boldsymbol{e}_i \times \boldsymbol{e}_j) \cdot \boldsymbol{e}_k$.

Exercise 1.7. Prove the following vector calculus identities in index notation

$$\operatorname{div} \operatorname{curl} \boldsymbol{A} = 0, \tag{1.234}$$

$$\operatorname{curl} \operatorname{grad} f = 0, \tag{1.235}$$

$$\operatorname{div}(\boldsymbol{A} \times \boldsymbol{B}) = \boldsymbol{B} \cdot \operatorname{curl} \boldsymbol{A} - \boldsymbol{A} \cdot \operatorname{curl} \boldsymbol{B}, \tag{1.236}$$

$$\operatorname{div}(f\boldsymbol{A}) = f \operatorname{div} \boldsymbol{A} + \boldsymbol{A} \cdot \operatorname{grad} f. \tag{1.237}$$

Are you convinced yet that the index notation is a good thing? If not, prove the above using the standard approach with explicit vector components.

Manifolds and Tensors

Exercise 1.8. Show that the Christoffel symbol transforms as follows:

$$\Gamma'^{n}_{ij} = \frac{\partial X'^{m}}{\partial X^{a}}\frac{\partial X^{b}}{\partial X'^{i}}\frac{\partial X^{c}}{\partial X'^{j}}\Gamma^{a}_{bc} - \frac{\partial X^{a}}{\partial X'^{i}}\frac{\partial X^{b}}{\partial X'^{j}}\frac{\partial^{2}X'^{m}}{\partial X^{a}\partial X^{b}}. \qquad (1.238)$$

Exercise 1.9. Show that $\Gamma^{n}_{ij}A^{j}$ transforms as follows:

$$\Gamma'^{n}_{ij}A'^{j} = \frac{\partial X'^{m}}{\partial X^{a}}\frac{\partial X^{b}}{\partial X'^{i}}\Gamma^{a}_{bc}A^{c} - \frac{\partial X^{b}}{\partial X'^{i}}\frac{\partial^{2}X'^{m}}{\partial X^{b}\partial X^{c}}A^{c}.$$

Exercise 1.10. Rewrite the line element of the hyperbolic disk in polar coordinates, this should give

$$ds^{2} = 4\frac{dr^{2} + r^{2}d\varphi^{2}}{(1 - r^{2})^{2}}.$$

Exercise 1.11 (takes time). Following on from the previous exercise, introduce the new radial coordinate $\rho = 2r/(1 - r^{2})$ and show that the line element becomes

$$ds^{2} = \frac{d\rho^{2}}{1 + \rho^{2}} + \rho^{2}d\varphi^{2}.$$

Lastly, introduce the hyperbolic 'angle' χ by using the new coordinate $\rho = \sinh\chi$ and show that we can now write

$$ds^{2} = d\chi^{2} + \sinh^{2}\chi d\varphi^{2}.$$

This somewhat motivates the name hyperbolic disk as this line element is very similar to Eq. (1.58), with the trigonometric function changed to the hyperbolic one.

Exercise 1.12. Show that the covariant divergence of the vector A^{i} can be written as

$$\nabla_{i}A^{i} = \frac{1}{\sqrt{-g}}\partial_{i}\left(\sqrt{-g}A^{i}\right).$$

Exercise 1.13 (takes time). Consider the line-element

$$ds^2 = dy^2 + \frac{dz^2}{\cosh^4 z}.$$

First, compute the Christoffel symbol components using Eq. (1.72), and second, find them using the geodesic equations obtained from the Euler–Lagrange equations of Eq. (1.74). Next, solve the geodesic equations and verify that they are straight lines. Find the coordinate transformation to Cartesian coordinates.

Exercise 1.14. Follow Theorem 1.1 and show that the connection in a space with torsion is given by

$$\Gamma^k_{ij} = \{^{\ k}_{ij}\} + T_{ij}{}^k - T_j{}^k{}_i + T^k{}_{ij},$$

where Γ^k_{ij} is the usual Christoffel symbol and $T_{ij}{}^k$ is the torsion tensor.

Exercise 1.15. Consider a covariant derivative satisfying $\tilde{\nabla}_a g_{bc} = Q_a g_{bc}$. Show that this implies $\tilde{\nabla}_a g^{bc} = -Q_a g^{bc}$.

Curvature

Exercise 1.16 (takes time). We began Sec. 1.3 by discussing parallel transport on the surface of a sphere. This exercise is the real thing: Recall the line-element of the surface of the 2-sphere $ds^2 = R^2(d\theta^2 + \sin^2\theta d\phi^2)$, for simplicity we will set $R = 1$.

(i) Solve the equations of parallel transport along constant longitudes, also called meridians. This means $\phi = \phi_0$ for such curves.

(ii) Solve the equations of parallel transport along constant latitudes sometimes called circles of latitude. This means $\theta = \theta_0$ for such curves.

(iii) Parallelly transport the vector $V^i = (1,0)$, from $(\pi/2, 0)$ to the North pole, then to $(\pi/2, \pi/2)$, and lastly back to the starting point $(\pi/2, 0)$.

(iv) Find the angle between V^i_{initial} and V^i_{final}.

Exercise 1.17. Consider the line element $ds^2 = v^2 du^2 + u^2 dv^2$. Show that this space has vanishing Riemann curvature tensor. Recall, it is sufficient to compute R_{1212}.

Exercise 1.18. In the previous exercise we showed that $ds^2 = v^2 du^2 + u^2 dv^2$ is Euclidean space in awkward coordinates. Hence, there must exist a coordinate transformation such that we can write this line element as $ds^2 = dx^2 + dy^2$. Find this coordinate transformation. *Note: This was a prize question set by Peter Hogan from University College Dublin when I was learning General Relativity. My prize was a copy of 'Gems of Hubble' by Mitton and Maran!*

Exercise 1.19 (hard). Prove theorem 1.5.

Exercise 1.20. Show that in n dimensions the Riemann curvature tensor has $n^2(n^2 - 1)/12$ independent components.

2

Einstein Field Equations

The aim of this chapter is to formulate Einstein's theory of General Relativity. While it is easy to state the field equations, it is much harder combining the physics and mathematics involved to motivate this theory. In doing so, we will also encounter different theories of gravity which can be viewed as extensions of General Relativity. As a starting point to this chapter we should state the main working hypothesis: The framework of differential geometry is suitable to formulate a consistent and physically meaningful theory of gravity. It is not clear whether this is possible or not. At this point this is a huge leap of faith and it might not work out. The idea of geometrising the different forces in nature has motivated a large amount of research.

2.1. Some Physics Background

2.1.1. *Newton's theory of gravity*

Newton's law of universal gravitational attraction is usually written in the form

$$F = -G\frac{Mm}{r^2}, \tag{2.1}$$

where M and m are two idealised point masses and $r^2 = x^2 + y^2 + z^2$ is the Euclidean distance between those masses. G is Newton's gravitational constant. Since the force F is a vector, forces have directions, one should write more correctly

$$\boldsymbol{F}_{12} = -G\frac{m_1 m_2}{|\boldsymbol{r}_{12}|^2}\hat{\boldsymbol{r}}_{12}, \tag{2.2}$$

where $r_{12} = r_2 - r_1$ is the vector pointing from the position of the first mass to the second mass. We will now choose r_1 as the origin of our coordinate system, we also set $M = m_1$ and $m = m_2$ to simplify the following discussion. It turns out that Newton's law of gravity is also valid for extended objects and their gravitational fields behave as if their masses were confined to an infinitesimal area at their centre of mass.

Newton's second law states $\boldsymbol{F} = m\boldsymbol{g}$ and hence the force per unit mass, or gravitational acceleration, can be written in either of two ways

$$m\boldsymbol{g} = \boldsymbol{F} = -G\frac{Mm}{|r|^2}\hat{\boldsymbol{r}}, \tag{2.3}$$

$$\text{or} \quad \boldsymbol{g} = \frac{1}{m}\boldsymbol{F} = -G\frac{M}{|r|^2}\hat{\boldsymbol{r}}. \tag{2.4}$$

Here r is the Euclidean distance from the origin. We can introduce the gravitational potential V because the gravitational force is conservative. Recall that this means that the work done by gravity from one point to another is path-independent or, more mathematically, the force vector is curl free. The Newtonian potential ϕ_N is defined via the equation

$$\boldsymbol{g} = -\operatorname{grad}\phi_N, \tag{2.5}$$

which means the gravitational potential is given by

$$\phi_N = -G\frac{M}{|r|}. \tag{2.6}$$

The subscript in ϕ_N indicates the Newtonian potential.

Relation (2.5) shows that the gravitational acceleration of a test mass is independent of the amount of mass which is accelerated. This observation goes back to Galileo and was well known before Newton formulated his force law. However, in principle Newton's theory would also be valid if we could distinguish between inertial mass, the m on the left-hand side of (2.3) and the gravitational mass,

m on the right-hand side of (2.3). The gravitational mass is the analogue of the electric charge in electromagnetism. In Newtonian gravity, the equivalence between inertial mass and gravitational mass is a fact deduced by experiments. It turns out that in General Relativity this fact is a necessary part of the theory and not an additional ingredient.

Equation (2.4) also provides us with a conceptual question: Can we, in principle, distinguish being accelerated by a powerful rocket from being in a gravitational field with identical magnitude? Einstein's equivalence principle states that we cannot distinguish these scenarios, this means we assume the physical equivalence of an accelerated reference frame and a gravitational field.

One can integrate Eq. (2.4) over the volume V with boundary ∂V which encloses the mass and arrive at Gauss' law of gravity

$$\oint_{\partial V} \boldsymbol{g} \cdot d\boldsymbol{S} = -4\pi GM. \tag{2.7}$$

In differential form this yields Poisson's equation for the gravitational field

$$\Delta \phi_{\mathrm{N}} = 4\pi G\rho, \tag{2.8}$$

where ρ is the density distribution of matter and one can define the mass as the volume integral of the density $M = \int_V \rho\, dV$. The Laplacian is denoted by Δ, in Euclidean coordinates it reads $\Delta = \partial_{xx} + \partial_{yy} + \partial_{zz}$.

For a spherically symmetric source with constant density $\rho = \rho_0 = \mathrm{const.}$ we choose $\phi = \phi(r)$ and have

$$\Delta \phi_{\mathrm{N}} = \frac{1}{r^2} \frac{d}{dr}\left(r^2 \frac{d\phi}{dr}\right) = 4\pi G\rho_0. \tag{2.9}$$

This leads to the well-known result

$$\phi_{\mathrm{N}} = -G\frac{M}{r}. \tag{2.10}$$

2.1.2.   *Special relativity*

All pre-relativistic theories of physics were based on the idea of a universal time which could be used in experiments. From a geometrical point of view, this means we were working in a flat 3D space with an external clock. The key idea of special relativity is to incorporate time into a geometrical framework which yields a 4D spacetime and formulate all physical theories within this framework.

One way to argue that this is indeed necessary comes from the speed of light c and our observations that no massive object can be accelerated to the point where it travels at the speed of light. If we accept that the speed of light is an upper speed limit, then we are immediately led to the following thought experiment. Consider a train travelling at speed $3c/4$ with a powerful rocket launcher in one of the carriages which can fire a rocket at $3c/4$. According to the pre-relativistic Galilean principle of relativity an observer at rest should observe the rocket to have a speed larger than the speed of light. However, this would violate our assumption that the speed of light is an upper limit and hence we are forced to reconsider what exactly happens in this kind of set up. Following this through yields the concept of time dilation which is the difference of the elapsed time of two events as seen by two observers in relative motion. In simple words, clocks tick at different rates depending on their speed. Clearly, this is in gross contradiction to the idea of a universal time. An event in special relativity is a point p in Minkowski space which is characterised by four coordinates, the time and three spatial coordinates.

Historically, the first clues of special relativity came out of Maxwell's equations which we will briefly discuss in the next section. The Maxwell equations are invariant under a particular transformation of the time and the space coordinates, the so-called Lorentz transformations. It was works of Poincaré, Einstein and Minkowski in the early 1900s which made the connection between the constancy of the speed of light and geometry.

When considering Euclidean 3-space $\mathbb{E}^3$, one notices that this space is invariant under rotations and under translations. Similarly, Minkowski space with coordinates $X^i = \{ct, x, y, z\}$

$$ds^2 = \eta_{ab}dX^a dX^b = -c^2 dt^2 + dx^2 + dy^2 + dz^2, \tag{2.11}$$

is also flat and should have some symmetries. In this case we can make 4D rotations and translations along each of the axes. It is these 4D rotations which are called Lorentz transformations. This means we can find matrices $L^i{}_j$ such that

$$\eta_{ab} = L^c{}_a L^d{}_b \eta_{cd}, \tag{2.12}$$

which means that they leave the Minkowski metric invariant. One can use Eq. (2.12) to derive the mathematical properties of the Lorentz transformations, for instance we can see immediately that $\det L^i{}_j = \pm 1$. The Lorentz transformations form a 6D group, the Lorentz group. However, as this is slightly outside the scope of this book we will only state the explicit form of boosts along the x-direction which are given by

$$L^c{}_a = \begin{pmatrix} \gamma & -\gamma\beta & 0 & 0 \\ -\gamma\beta & \gamma & 0 & 0 \\ 0 & 0 & 1 & 0 \\ 0 & 0 & 0 & 1 \end{pmatrix}, \tag{2.13}$$

where $\beta = v/c$ is the velocity normalised by the speed of light, and $\gamma = 1/\sqrt{1 - \beta^2}$ is the well-known Lorentz factor which is at the heart of special relativity. When applied to the coordinates directly we have $X'^a = L^a{}_b X^b$, so that the Lorentz transformations (2.13) are explicitly given by

$$ct' = \gamma(ct - \beta x), \tag{2.14}$$

$$x' = \gamma(x - \beta ct), \tag{2.15}$$

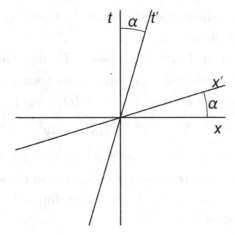

Fig. 2.1   Visualisation of the Lorentz transformations. The angle α is determined by the velocity β and is given by $\beta = \tan(\alpha)$, when $\beta \to 1$ which corresponds to the speed of light, then $\alpha = \pi/4$ which would correspond to the diagonals.

$$y' = y, \tag{2.16}$$

$$z' = z. \tag{2.17}$$

We can visualise these transformations in Fig. 2.1. The x'-axis corresponds to all simultaneous events for which $t' = 0$, similar to the x-axis which can be defined by $t = 0$. This means that the notion of simultaneity depends on the velocity of the observer and two observers may not agree on two events being simultaneous.

Probably the most important geometrical fact about Minkowski space is that it contains two distinct regions which are separated by the so-called light cone. The light cone is defined by $ds^2 = 0$ and corresponds to diagonal lines which indicate the path that light would take through spacetime. Two points (events) for which $ds^2 < 0$ are said to be separated by a time-like interval, in this case we will be able to define proper time. Physically speaking this means a particle with speed $v < c$ can travel from one event to the other and we can introduce a time order for these events. In the case of the light cone one speaks of null intervals or light-like intervals $ds^2 = 0$.

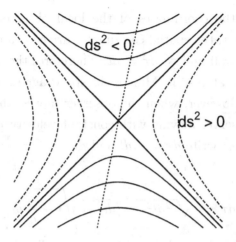

Fig. 2.2 Minkowski diagram. The diagonal lines correspond to the path travelled by light emitted at the origin, they correspond to null intervals $ds^2 = 0$. The dashed hyperbolas correspond to events separated by constant space-like intervals $ds^2 > 0$, the solid hyperbolas correspond to events separated by constant time-like intervals $ds^2 < 0$. The dotted straight line corresponds to a massive particle travelling with constant speed $v < c$. If we rotate this figure along the vertical axis, the diagonal lines would become a cone, hence the name light cone.

Last, when $ds^2 > 0$ one speaks of space-like intervals. This means that no signal can be exchanged between these two events, there is no causal relationship between them. One can visualise Minkowski space as in Fig. 2.2.

It is possible to formulate any physical theory within the setting of Minkowski space so that the physical equations respect Lorentz invariance. This means for instance that all propagation speeds should be smaller than the speed of light. Moreover, in the limit of small velocities we should always recover the physical theory in the standard Newtonian formulation.

The proper time seen by a local observer on its local clock is denoted by $d\tau$ and is related to the line element simply by $d\tau^2 = -ds^2/c^2$, which can also be written as

$$d\tau = \frac{1}{c}\sqrt{-\eta_{ab}dX^a dX^b}, \qquad (2.18)$$

where X^a are the coordinates of the local observer. If this local observer is at rest, the spatial coordinates are constant which means $dx = dy = dz = 0$ and so $d\tau = dt$. Therefore, the proper time of a local observer at rest in Minkowski space agrees with the coordinate time t. However, when the observer moves, this will change. Consider an observer moving with constant velocity v_x along the x-axis, then we can write $dx = v_x dt$ and $dy = dz = 0$ and the proper time becomes

$$d\tau = \frac{1}{c}\sqrt{c^2 dt^2 - v_x^2 dt^2} = \sqrt{1 - \frac{v_x^2}{c^2}}\, dt. \qquad (2.19)$$

Using the previously used notation with the Lorentz factor γ, this becomes the famous time dilation equation of special relativity

$$dt = \frac{1}{\sqrt{1 - \frac{v_x^2}{c^2}}} d\tau = \gamma d\tau. \qquad (2.20)$$

It means that the clock cycle depends on the relative velocity of the observer. For $0 < v_x < c$ the Lorentz factor is greater than one, $\gamma > 1$. Therefore, the clock cycle is increasing (more time between ticks) and hence, the moving clock appears to be running slower. For velocities much smaller than the speed of light, this effect is very small and hardly detectable. However, various experiments are in excellent agreement with the time dilation formula.

Similarly, let us consider a rod of length $l = x_2 - x_1$ at some fixed time, where x_1 and x_2 denote the rod's endpoints. We assume this rod to be moving along the positive x-direction. Now we boost along the x-direction using the speed of the rod and compute l'. Since the rod is at rest in the boosted coordinates, l' is the rod's length at rest. Equation (2.15) immediately implies that $l' = x_2' - x_1' = \gamma(x_2 - x_1) = \gamma l$. As before, for $0 < v_x < c$ we have $\gamma > 1$ and therefore $l < l'$ which implies that the moving rod is contracted. This effect is called length contraction.

Consider a massive point particle and assume its motion is described by the curve $X^i(\tau)$. The path through spacetime which is described by the curve $X^i(\tau)$ is called the world line of the particle. The tangent vector to this curve is $dX^i/d\tau$, when considering point particles in special relativity one often denotes this tangent vector by u^i calling it 4-velocity. Moreover, we use Eq. (2.20) to change the parameter τ to physical time t which gives

$$u^i = \frac{dX^i}{d\tau} = \gamma \frac{dX^i}{dt} = \gamma \begin{pmatrix} c \\ v^1 \\ v^2 \\ v^3 \end{pmatrix}. \tag{2.21}$$

In analogy to classical mechanics we define the 4-momentum by $p^i = mu^i$ and interpret $p^0 = E/c$ as the energy while the spatial components correspond to momentum. This yields the very well-known formula

$$E = \gamma mc^2 = \frac{mc^2}{\sqrt{1 - v^2/c^2}} = mc^2 + \frac{m}{2}v^2 + O(v^3/c^3), \tag{2.22}$$

which in lowest order is the mass–energy equivalence relation.

2.1.3. *Maxwell equations*

Using Euclidean 3-vectors the Maxwell equations are written using the electric field $\boldsymbol{E}$, magnetic field $\boldsymbol{B}$ with appropriate sources ρ and $\boldsymbol{j}$. The homogeneous Maxwell equations are

$$\operatorname{curl} \boldsymbol{E} + \frac{1}{c}\frac{\partial \boldsymbol{B}}{\partial t} = 0, \quad \operatorname{div} \boldsymbol{B} = 0, \tag{2.23}$$

and the inhomogeneous Maxwell equations are given by

$$\operatorname{div} \boldsymbol{E} = 4\pi\rho, \quad \operatorname{curl} \boldsymbol{B} - \frac{1}{c}\frac{\partial \boldsymbol{E}}{\partial t} = \frac{4\pi}{c}\boldsymbol{j}. \tag{2.24}$$

Here ρ is the charge density and $\boldsymbol{j}$ is the current density, and we work with Gaussian units. In addition to the Maxwell equations we also need to specify the force acting on a charged particle, this is the Lorentz force

$$\boldsymbol{F} = q \left(\boldsymbol{E} + \frac{\boldsymbol{v}}{c} \times \boldsymbol{B} \right). \tag{2.25}$$

Interestingly, in Gaussian units the electric and the magnetic field have the same dimensions, in physical SI units, they differ exactly by units of velocity. This is no coincidence and points toward special relativity.

It is also possible to formulate Maxwell's equations in Minkowski space with metric η_{ij}. We are working with coordinates $X^i = \{ct, x, y, z\}$, $i = 0, 1, 2, 3$. The Faraday tensor F_{ij} is defined by

$$F_{ij} = \begin{pmatrix} 0 & E_x & E_y & E_z \\ -E_x & 0 & -B_z & B_y \\ -E_y & B_z & 0 & -B_x \\ -E_z & -B_y & B_x & 0 \end{pmatrix}, \tag{2.26}$$

it is skew-symmetric $F_{ij} = -F_{ji}$. This means we can write $E_\alpha = F_{0\alpha}$ and $B_\alpha = -1/2\, \varepsilon_{\alpha\beta\gamma} F^{\beta\gamma}$. Recall that our time coordinate is $X^0 = ct$, therefore all components of the Faraday tensor have the same dimensions. If we were to work with coordinates $X^i = \{t, x, y, z\}$, we would have to be careful with factors of c in F_{ij} and the units of time and space.

The Faraday tensor can be expressed using a 4-vector potential $F_{ij} = \partial_i A_j - \partial_j A_i$ with $A_i = (-\phi, \boldsymbol{A})$. The magnetic field is defined by $\boldsymbol{B} = \operatorname{curl} \boldsymbol{A}$ and the electric field is $\boldsymbol{E} = -\operatorname{grad} \phi - \partial \boldsymbol{A}/\partial t$.

Let us recall Theorem 1.3 which implies that the Faraday tensor must satisfy the equation

$$\partial_i F_{jk} + \partial_k F_{ij} + \partial_j F_{ki} = 0. \tag{2.27}$$

These equations turn out to be equivalent with the homogeneous Maxwell equations (2.23). On the other hand, the inhomogeneous Maxwell equations become

$$\partial_j F^{ij} = -\frac{4\pi}{c} J^i,\tag{2.28}$$

where the 4-vector current has components $J^i = (c\rho, \boldsymbol{j})$. Due to the skew-symmetry of F^{ij}, the latter equation directly implies that the current is conserved. We have $\partial_i \partial_j F^{ij} = 0$ and therefore $\partial_i J^i = 0$ which is the continuity equation, or the (local) charge conservation equation.

Setting $i = 0$ in Eq. (2.28) leads to

$$\partial_1 F^{01} + \partial_2 F^{02} + \partial_3 F^{03} = -4\pi\rho,$$

$$\partial_1(-E_x) + \partial_2(-E_y) + \partial_3(-E_z) = -4\pi\rho,\tag{2.29}$$

which is equivalent to

$$\text{div}\, \boldsymbol{E} = 4\pi\rho.\tag{2.30}$$

Next, setting $i = 1$ in Eq. (2.28) we find

$$\partial_0 F^{10} + \partial_2 F^{12} + \partial_3 F^{13} = -\frac{4\pi}{c} j^1,$$

$$\frac{1}{c}\partial_t(E_x) + \partial_2(-B_z) + \partial_3(B_y) = -\frac{4\pi}{c} j^1.\tag{2.31}$$

This is the x-component of the equation

$$\text{curl}\, \boldsymbol{B} - \frac{1}{c}\frac{\partial \boldsymbol{E}}{\partial t} = \frac{4\pi}{c}\boldsymbol{j},\tag{2.32}$$

and likewise for the other components, hereby proving that Eq. (2.28) is indeed equivalent to Eq. (2.24).

Maxwell's equations not only provide a link between geometry and physics, they also contain the fundamental ideas of modern gauge field theories which we will briefly discuss. The Maxwell equations can be written compactly using the Faraday tensor which itself can be expressed using the 4-vector potential $F_{ij} = \partial_i A_j - \partial_j A_i$. If we change

A_i by the gradient of a scalar function χ, say, $A_i \to A_i + \partial_i \chi$, then the Faraday tensor will be invariant under this change since partial derivatives commute. This means that A_i contains a non-physical degree of freedom which we may eliminate or fix by choosing a specific gauge. One popular gauge in electromagnetism is the Lorenz (not Lorentz) gauge whereby one sets $\partial_i A^i = 0$. Another often-used gauge is the Coulomb gauge defined by div $\boldsymbol{A} = 0$.

2.1.4. *Matter tensors*

Let us begin with recalling the stress tensor or Cauchy stress tensor σ_{ij} of continuum mechanics. The geometrical setting is Euclidean 3-space with Cartesian coordinates. This is a rank-2 tensor with nine components and it defines the state of stress at any point inside a deformed material. The diagonal components of this tensor are usually called the normal stresses, while the remaining components are called shear·stresses. For a hydrostatic fluid in equilibrium the stress tensor is given by $-p\delta_{ij}$ with p being the hydrostatic pressure, pressure is used instead of stress when the material is compressible.

Since all known physical theories can be formulated within the framework of special relativity, the stress tensor is superseded by the energy–momentum tensor (sometimes called stress–energy tensor or stress–energy–momentum tensor) T^{ij}, also a rank-2 tensor but in Minkowski space. In addition to the stresses it contains information about the energy density and energy fluxes of the matter. The energy–momentum tensor satisfies the four conservation equations $\partial_i T^{ij} = 0$. We can interpret the $j = 0$ equation as the energy conservation equation, and the three equations $j = 1, 2, 3$ as the momentum conservation equations.

For an ideal fluid in thermodynamic equilibrium the energy–momentum tensor in the fluid's reference frame is given by

$$T^{ij} = \begin{pmatrix} \rho c^2 & 0 & 0 & 0 \\ 0 & p & 0 & 0 \\ 0 & 0 & p & 0 \\ 0 & 0 & 0 & p \end{pmatrix}. \tag{2.33}$$

The hydrostatic pressure is p and ρ denotes the fluid's energy density.

As before, we work with coordinates $X^i = \{ct, x, y, z\}$ in Minkowski space $\eta_{ij} = \mathrm{diag}(-1, 1, 1, 1)$. Let u^a be a unit time-like vector $\eta_{ij}u^i u^j = -1$ representing the velocity of the fluid, then we can write the energy–momentum tensor as

$$T^{ij} = (\rho c^2 + p)u^i u^j + p\eta^{ij} = \rho c^2 u^i u^j + p(\eta^{ij} + u^i u^j), \tag{2.34}$$

so that we have $T^{ij}u_i u_j = \rho c^2$. In the fluid's rest frame we simply have $u^i = (1, 0, 0, 0)$.

The latter formulation used in Eq. (2.34) is geometrically useful. Let us define the tensor $q^i_j = \delta^i_j + u^i u_j$, then we can immediately verify the following relations

$$\begin{aligned} q^i_j q^j_k &= (\delta^i_j + u^i u_j)(\delta^j_k + u^j u_k) \\ &= \delta^i_k + u^i u_k + u^i u_k + u^i u_j u^j u_k \\ &= \delta^i_k + u^i u_k = q^i_k \end{aligned} \tag{2.35}$$

and also

$$q^i_j u^j = (\delta^i_j + u^i u_j)u^j = u^i + u^i u_j u^j = 0. \tag{2.36}$$

Hence the tensor q^i_j has a natural interpretation in terms of projections, namely it projects onto the surface normal to the vector u^j.

Projections along a unit normal are frequently used in General Relativity, especially in the initial value formulation.

Let us have a closer look at the conservation equation $\partial_i T^{ij} = 0$ for the perfect fluid in Minkowski space. We note that $u_i u^i = -1$ implies $u_i \partial_j u^i = 0$. Written out explicitly, the conservation equation is given by

$$(\partial_i \rho c^2) u^i u^j + \rho c^2 (\partial_i u^i) u^j + \rho c^2 u^i (\partial_i u^j)$$
$$+ (\partial_i p)(\eta^{ij} + u^i u^j) + p(\partial_i u^i) u^j + p u^i (\partial_i u^j) = 0. \quad (2.37)$$

We will decompose these j equations into two parts. First, we consider $-u_j \partial_i T^{ij} = 0$ which becomes

$$(\partial_i \rho c^2) u^i + \rho c^2 (\partial_i u^i) + p(\partial_i u^i) = 0. \quad (2.38)$$

Second, we consider $q_j^k \partial_i T^{ij}$. Here we find

$$(\rho c^2 + p) u^i \partial_i u^k + (\partial_i p)(\eta^{ik} + u^i u^k) = 0. \quad (2.39)$$

Next, we will consider the non-relativistic limit of those equations. In this limit $p/c^2 \ll \rho$ and we also assume that velocities are small when compared to the speed of light. The 4-velocity is given by $u^i = (1, \boldsymbol{v}/c) = (1, v_x/c, v_y/c, v_z/c)$ and this means we will assume $|\boldsymbol{v}| \ll c$. Therefore,

$$u^i \partial_i \rho = \frac{1}{c} \left[\frac{\partial}{\partial t} \rho + \boldsymbol{v} \cdot \nabla \rho \right], \quad (2.40)$$

$$\partial_i u^i = \frac{1}{c} \nabla \cdot \boldsymbol{v}, \quad (2.41)$$

$$u^i \partial_i u^k = \frac{1}{c^2} \left[\frac{\partial}{\partial t} \boldsymbol{v} + (\boldsymbol{v} \cdot \nabla) \boldsymbol{v} \right]. \quad (2.42)$$

So Eq. (2.38) can now be written as follows:

$$\frac{\partial}{\partial t} \rho + \boldsymbol{v} \cdot \nabla \rho + (\nabla \cdot \boldsymbol{v})(\rho + p/c^2) = 0. \quad (2.43)$$

We will take into account $p/c^2 \ll \rho$ and arrive at

$$\frac{\partial}{\partial t} \rho + \nabla \cdot (\rho \boldsymbol{v}) = 0. \quad (2.44)$$

This is the well-known continuity equation of fluid dynamics. It simply means that the flow of matter into the system equals the flow out of the system, provided there are no sinks or sources. This is equivalent to the statement that energy of this system is conserved.

Next, let us consider Eq. (2.39) which becomes

$$(\rho+p/c^2)\left[\frac{\partial}{\partial t}\boldsymbol{v} + (\boldsymbol{v}\cdot\nabla)\boldsymbol{v}\right] + \nabla p + \frac{1}{c^2}\left[\frac{\partial}{\partial t}p + \boldsymbol{v}\cdot\nabla p\right]\boldsymbol{v} = 0. \quad (2.45)$$

We assume that pressure changes over time are slow relative to the speed of light, this means $(|\boldsymbol{v}|/c)(\partial p/c\partial t) \ll \nabla p$. Then we arrive at

$$\left[\frac{\partial}{\partial t}\boldsymbol{v} + (\boldsymbol{v}\cdot\nabla)\boldsymbol{v}\right]\rho = -\nabla p. \quad (2.46)$$

This second equation is the Euler equation of fluid dynamics, which corresponds to the conservation of momentum. Both these two little results are very nice as they show how non-relativistic physics naturally emerges from the relativistic treatment.

We have already discussed Maxwell's equations and will now mention the energy–momentum tensor of the electromagnetic field. Let us consider the Maxwell equations (2.28) and apply F_{ik} to both sides. This gives

$$-\frac{4\pi}{c}j^i F_{ik} = F_{ik}\partial_j F^{ij} = \partial_j(F^{ij}F_{ik}) - F^{ij}\partial_j F_{ik}. \quad (2.47)$$

Using Eq. (2.27) we can write

$$F^{ij}\partial_k F_{ij} = 2F^{ij}\partial_j F_{ik}, \quad (2.48)$$

next re-write the right-hand side of Eq. (2.47) in the following way

$$-\frac{4\pi}{c}j^i F_{ik} = \partial_j(F^{ij}F_{ik}) - F^{ij}\partial_j F_{ik}$$

$$= \partial_j(F^{ij}F_{ik}) - \frac{1}{2}F^{ij}\partial_k F_{ij}$$

$$= \partial_j (F^{ij} F_{ik}) - \frac{1}{4} \partial_k (F^{mn} F_{mn})$$

$$= \partial_j \left[F^{ij} F_{ik} - \frac{1}{4} \delta_k^j F^{mn} F_{mn} \right]. \tag{2.49}$$

The term in the square brackets is the energy–momentum tensor of the electromagnetic field

$$T_k^j = \frac{1}{4\pi} \left(F^{ij} F_{ki} + \frac{1}{4} \delta_k^j F^{mn} F_{mn} \right), \tag{2.50}$$

which satisfies the equation $\partial_i T_k^i = -j^i F_{ik}$. This means that in the absence of charges and current $j^i = 0$, the energy–momentum tensor is conserved and satisfies the usual equation $\partial_j T_k^j = 0$. In the presence of sources, the electromagnetic field is not conserved, however, the total energy–momentum tensor of the field plus the sources are conserved. The equation $\partial_j T_k^j = -j^i F_{ik}$ means that any deviation from the conservation of T_k^j is due the sources j^i.

For concreteness let us compute T^{00} explicitly, for which we find

$$T^{00} = \frac{1}{8\pi} \left(|\boldsymbol{E}|^2 + |\boldsymbol{B}|^2 \right). \tag{2.51}$$

As expected, this is indeed the energy–density of the electromagnetic field.

2.2.   Geometry and Gravity

2.2.1.   *Geodesics and Newton's law*

Newton's second law of motion in the context of gravitational fields states

$$\frac{d^2 \boldsymbol{x}}{dt^2} = \boldsymbol{g}. \tag{2.52}$$

Let us compare this equation with the geodesic equation (1.70) discussed in Chap. 1. This was given by

$$\frac{d^2 x^a}{d\lambda^2} = -\Gamma_{bc}^a \frac{dx^b}{d\lambda} \frac{dx^c}{d\lambda}. \tag{2.53}$$

If it is possible to describe gravity using the language of differential geometry (our working hypothesis) then we should identify the Christoffel symbols in the geodesic equation with the force per unit mass in Newton's second law. Let us take into account physical units. We would assign units of length, m for meters, to our spatial coordinates, we measure time in seconds s, and masses in kg, using SI units. We denote the physical units of a quantity by square brackets. The parameter λ in the geodesic equation (2.53) is dimensionless. We have $[\boldsymbol{g}] = \text{m/s}^2$ and find that $[\Gamma] = 1/\text{m}$. Hence our first observation is that

$$\frac{[\boldsymbol{g}]}{[\Gamma]} = \text{m}^2/\text{s}^2 = (\text{m/s})^2, \tag{2.54}$$

which has the units of velocity squared. The only fundamental physical constant with such units is the speed of light c which is of paramount importance in modern theoretical physics and is the building block of special relativity. Hence, we will make the following identification

$$\frac{1}{c^2}\boldsymbol{g} \quad \longleftrightarrow \quad \Gamma^a_{bc}. \tag{2.55}$$

Let us recall that the Christoffel symbol components Γ^a_{bc} depend on the metric and, more importantly, on its first partial derivatives $\partial_k g_{ij}$. On the other hand, in the Newtonian theory $\boldsymbol{g} = -\operatorname{grad}\phi_{\text{N}}$, so the field depends on the derivatives of the gravitation potential. This would suggest that the metric components g_{ij} contain the gravitational potential ϕ_{N} which means

$$\frac{1}{c^2}\phi_{\text{N}} \quad \longleftrightarrow \quad \frac{1}{2}g_{ij}, \tag{2.56}$$

where we recall the factor of 1/2 in the definition of the Christoffel symbol. A direct consequence of this identification is that under the assumption of a static and spherically symmetric gravitational field,

Eq. (2.10), we would expect

$$\frac{2GM}{c^2 r} \quad \longleftrightarrow \quad g_{ij}. \tag{2.57}$$

We can improve this slightly by considering the limit $M \to 0$ in which case we would expect to find an empty space whose metric is flat, in particular we would expect to find Minkowski space, see Eq. (1.60), the space of special relativity. Therefore, up to an overall sign, we arrive at the identification

$$1 \pm \frac{2GM}{c^2 r} \quad \longleftrightarrow \quad g_{ij}. \tag{2.58}$$

We will now try to make this identification more precise. Start with coordinates $X^i = \{ct, x, y, x\}$. Let us consider the static line element

$$ds^2 = -(1 + 2\phi)c^2 dt^2 + (1 - 2\phi)\left[dx^2 + dy^2 + dz^2\right], \tag{2.59}$$

assuming that $\phi = \phi(x, y, z) \ll 1$, and consider the motion of a freely falling massive particle with 4-velocity $u^i = dX^i/d\lambda$. The trajectory of the particle is determined by the geodesic equation and satisfies

$$\dot{u}^i + \Gamma^i_{mn} u^m u^n = 0. \tag{2.60}$$

For non-relativistic motion $u^0 \gg u^1, u^2, u^3$ so that

$$\dot{u}^i + \Gamma^i_{00} u^0 u^0 = 0. \tag{2.61}$$

In lowest order in ϕ one finds that $\Gamma^0_{00} = 0$, and therefore the $i = 0$ equation simplifies to $du^0/d\lambda = 0$ or $u^0 = C = \text{const}$. Since $u^0 = cdt/d\lambda$ we can express the geodesic parameter in terms of time as

$$\frac{cdt}{d\lambda} = C, \quad cdt = Cd\lambda. \tag{2.62}$$

The geodesic equation becomes

$$\dot{u}^\alpha + \Gamma^\alpha_{00} u^0 u^0 = 0, \tag{2.63}$$

and we can change the independent variable λ to t which results in

$$\frac{d^2 X^\alpha}{dt^2} = -\Gamma^\alpha_{00} c^2. \tag{2.64}$$

We note that the component Γ^α_{00} of Eq. (2.59) is given by $\Gamma^\alpha_{00} = \partial^i \phi$ which leads to

$$\frac{d^2 X^\alpha}{dt^2} = -\partial^\alpha \phi c^2, \tag{2.65}$$

which is consistent with Eq. (2.52). In turn, this establishes that the metric defined by Eq. (2.59) correctly reproduces Newton's law, taking into account our various approximations. Next, we must take a closer look at curvature with the aim of getting the Poisson equation out of geometry. We can see that ϕ in Eq. (2.52) is the correct Newtonian gravitational potential with an additional factor containing the speed of light

$$\phi = -\frac{GM}{c^2 r} = \frac{1}{c^2} \phi_N, \tag{2.66}$$

which is also in agreement with Eq. (2.58).

2.2.2. *Curvature and the Poisson equation*

Next we must address the question of finding the geometric analogue of the Poisson equation. Let us begin by recalling the geodesic deviation equation (1.213) given by

$$\frac{D^2 N^a}{D\lambda^2} = -(R_{jic}{}^a T^j T^c) N^i, \tag{2.67}$$

with the first two indices interchanged. Let us derive an analogous equation in the setting of Newtonian mechanics in the presence of a gravitational field. We should begin with Euclidean space with Cartesian coordinates $X^\alpha = \{x, y, z\}$, $\alpha = 1, 2, 3$ and a family of curves $X^\alpha(t, s)$ where t is time along a curve and s labels the curves. The tangent vector to any such curve is given by $T^\alpha = dX^\alpha / dt$ for fixed value s_0. The vector connecting infinitesimally close curves is $N^\alpha = dX^\alpha / ds$ for fixed times. The rate of change of N^α with respect to time can be interpreted as the relative velocity between nearby curves and its derivative as the relative acceleration. This is the quantity we wish to find, in analogy to Eq. (1.209). First, we have

$$\frac{\partial}{\partial s}V^\alpha = \frac{\partial}{\partial t}N^\alpha, \tag{2.68}$$

and the relative acceleration is given by

$$\frac{\partial^2}{\partial t^2}N^\alpha = \frac{\partial}{\partial t}\frac{\partial}{\partial s}V^\alpha = \frac{\partial}{\partial s}\frac{\partial}{\partial t}V^\alpha$$

$$= \frac{\partial}{\partial s}\left(\frac{\partial^2}{\partial t^2}X^\alpha\right) = -\frac{\partial}{\partial s}\left(\partial^\alpha\phi\right). \tag{2.69}$$

where we used Newton's second law. Next, we apply the chain rule to the last term

$$\frac{\partial}{\partial s}\left(\frac{\partial}{\partial X^\alpha}\phi\right) = \frac{\partial^2\phi}{\partial X^\alpha\partial X^\beta}\frac{\partial X^\beta}{\partial s} = \frac{\partial^2\phi}{\partial X^\alpha\partial X^\beta}N^\beta. \tag{2.70}$$

Therefore, we arrive at the Newtonian equivalent of the geodesic deviation equation

$$\frac{\partial^2 N^\alpha}{\partial t^2} = -\left(\partial_\alpha\partial_\beta\phi\right)N^\beta. \tag{2.71}$$

At this point we can change the indices α, β to full spacetime indices a, b by adding trivial components wherever necessary. This naturally leads to the identification

$$R_{jic}{}^a T^j T^c \quad \longleftrightarrow \quad \partial^a\partial_i\phi. \tag{2.72}$$

In view of Eq. (2.8), we would like to sum over the indices a and i so that the right-hand side becomes the Laplacian, provided that ϕ is time-independent. Using $R_{jic}{}^i T^j T^c = R_{jc}T^j T^c$ leads to

$$R_{jc}T^j T^c \quad \longleftrightarrow \quad \partial^i\partial_i\phi. \tag{2.73}$$

This identification is not unexpected since the Riemann curvature tensor contains the second partial derivatives of the metric and we already established that the metric should contain the gravitational potential. Hence Eq. (2.73) is consistent with our previous discussion which led to Eq. (2.58).

Consequently, in the absence of any gravitational fields, nearby geodesics should experience no relative acceleration and therefore we can conjecture that the vacuum field equations should take the form $R_{jc}T^jT^c = 0$ for all tangent vectors T^a which is equivalent to $R_{ij} = 0$. This leads to the our final identification

$$R_{ij} = 0 \quad \longleftrightarrow \quad \partial^a\partial_a\phi = 0. \tag{2.74}$$

Surprisingly, these are the correct vacuum field equations of General Relativity. For instance, the Minkowski metric (1.60) satisfies this equation. In order to discuss the inclusion of matter, we need to recall our discussion of matter tensors.

2.2.3.   Field equations of General Relativity

By comparing the geodesic deviation equation in a Lorentzian space with its equivalent in the Newtonian setting, we arrived at a sensible guess for the form of the vacuum field equations of General Relativity. In our discussion of matter tensors, we encountered the important energy–momentum conservation equation $\nabla_a T^{ab} = 0$. It seems natural to place the energy–momentum tensor onto the right-hand side of the Einstein field equations, in analogy to Eq. (2.8), with an appropriate coupling constant which should also contain Newton's gravitational constant. However, this also suggests that the left-hand side of the Einstein field equations cannot be based on the Ricci tensor alone, since $\nabla_a R^{ab} \neq 0$. Einstein indeed originally proposed the field equations $R_{ab} = \kappa T_{ab}$, however, this was soon dismissed due to the mentioned inconsistency. We require an object which contains the Ricci tensor and whose covariant divergence vanishes. The key lies in the twice contracted Bianchi identities, see Theorem 1.6. This shows that $G_{ab} = R_{ab} - Rg_{ab}/2$ has the desired property which justifies the name Einstein tensor for this object.

In November 1915, Einstein proposed the gravitational field equations

$$R_{ij} - \frac{1}{2}Rg_{ij} = \kappa T_{ij}, \quad \text{or} \quad G_{ij} = \kappa T_{ij}, \tag{2.75}$$

where κ is a coupling constant which needs to be related to Newton's gravitational constant.

Let us have a closer look at these field equations. Applying g^{ij} to both sides gives $-R = \kappa T$ where $T = g^{ij}T_{ij}$ is the trace of the energy–momentum tensor, and we needed to use $g^{ij}g_{ij} = 4$. Substituting R for T in Eq. (2.75) gives us an alternative form of the field equations, namely

$$R_{ij} = \kappa \left(T_{ij} - \frac{1}{2}Tg_{ij} \right). \tag{2.76}$$

In the absence of matter $T_{ij} = 0$ and the vacuum field equations are simply $R_{ij} = 0$. For this reason one often refers to metrics satisfying the vacuum field equations as Ricci flat. Note that this does not imply that the Riemann curvature vanishes. In fact, there are many known Ricci flat metrics with a singular Riemann tensor, one of which is the Schwarzschild metric which we will study in detail in Secs. 3.2 and 3.6.

Since the Ricci tensor, the metric tensor and the energy–momentum tensor are all symmetric tensors, the field equations (2.75) are a set of 10 nonlinear coupled partial differential equations in the components of the metric g_{ij}, the sources are given by the energy–momentum tensor. In n dimensions we would have $n(n + 1)/2$ equations, however, all 2D metrics satisfy $G_{ab} = 0$. It should be noted that we can make arbitrary coordinate transformations which can change the metric, since there are four coordinates we expect the number of independent equations to be reduce by four. Therefore, we are led to believe that the Einstein field equations are six independent equations plus four gauge fixing degrees of freedom.

Shortly after the formulation of the field equations (2.75) Einstein was interested in a particular solution, now known as the Einstein static universe, see Sec. 4.2.3. It is a cosmological solution of the field equations which corresponds to a universe without any dynamics. It turns out that the field equations do not allow for such a solution if the source is a perfect fluid. This motivated the introduction of the so-called cosmological constant Λ into the field equations which then take the form

$$R_{ij} - \frac{1}{2}Rg_{ij} + \Lambda g_{ij} = \kappa T_{ij}, \quad \text{or} \quad G_{ij} + \Lambda g_{ij} = \kappa T_{ij}. \qquad (2.77)$$

These field equations allow for such a static solution. One verifies that the left-hand side of the cosmological field equations still has vanishing covariant divergence since Λ is assumed to be a constant and the metric has vanishing covariant derivative. Sometimes Eq. (2.77) are referred to as the cosmological Einstein field equations. The physical units of the cosmological constant are $[\Lambda] = 1/\text{m}^2$ so that $1/\sqrt{\Lambda}$ is a characteristic length scale. The cosmological constant will be discussed in greater detail in Chap. 4.

2.2.4. *The principle of minimal gravitational coupling*

The main idea of special relativity was to reformulate physical theories in a flat 4D spacetime. General Relativity, on the other hand, is based on a curved manifold. Therefore, we need to address the question of how to make special relativistic theories compatible with General Relativity.

One way to achieve this is by simply replacing the Minkowski metric by a general metric $\eta_{ij} \rightarrow g_{ij}$, and by replacing all partial derivatives with covariant derivatives $\partial_i \rightarrow \nabla_i$. Sometimes this is called the principle of covariance. This also means that physical laws take the same form in all coordinate systems which is also referred to as the general principle of relativity. More technically speaking the theory is diffeomorphism invariant, this means invariant under general coordinate transformations.

However, this procedure alone does not result in a unique theory. This can be seen as follows. Assume we have a theory which contains a term of the form $\partial_i \partial_j v_k$, then following the above rule would give $\nabla_i \nabla_j v_k$. On the other hand, had we started with $\partial_j \partial_i v_k$ then we would arrive at $\nabla_j \nabla_i v_k$. In Minkowski space our partial derivatives commute and there is no preferred order, however, on arbitrary manifolds our two suggested terms differ exactly by the Riemann curvature tensor, see Eq. (1.165).

This is particularly relevant for Maxwell's equations, as we will see in the following. Let us work with the 4-vector potential A_i in the Lorenz gauge $\nabla_i A^i = 0$. Begin with Eq. (2.28) which gives

$$\nabla_j(\nabla^i A^j - g^{jk}\nabla_k A^i) = \nabla_j \nabla^i A^j - g^{jk}\nabla_j \nabla_k A^i$$
$$= \nabla^i(\nabla_j A^j) + R^i{}_k A^k - g^{jk}\nabla_j \nabla_k A^i$$
$$= R^i{}_k A^k - g^{jk}\nabla_j \nabla_k A^i \qquad (2.78)$$

and we arrive at

$$\nabla^j \nabla_j A^i - R^i{}_j A^j = \frac{4\pi}{c} J^i. \qquad (2.79)$$

On the other hand, we could have started with $\partial_j(\partial^i A^j - \partial^j A^i) = -\partial_j \partial^j A^i$, using the Lorenz gauge condition. Then, replacing partial with covariant derivatives would have given

$$\nabla^j \nabla_j A^i = \frac{4\pi}{c} J^i, \qquad (2.80)$$

without the Ricci tensor part. Equations (2.79) and (2.80) are different which shows that the principle of minimal coupling is not unique. Some additional input is needed in order to determine which version of Maxwell's equations should be used. In typical laboratory settings, the gravitational fields are weak and we will not be able to distinguish the two possible theories. However, the continuity equation of the 4-vector current becomes $\nabla_i J^i = 0$, using the principle of minimal coupling. Therefore, we can check the consistency of the proposed equations with the charge conservation equation which we assume to hold. One needs to apply ∇_i to (2.79) and (2.80), and in both cases one needs to interchange the order of the covariant derivatives, hereby introducing curvature terms.

We have

$$
\begin{aligned}
\nabla_i \nabla^j \nabla_j A^i &= \nabla^j \nabla_i \nabla_j A^i - R_i{}^s \nabla_s A^i + R^j{}_s \nabla_j A^s \\
&= \nabla^j \left[\nabla_j \nabla_i A^i + R_{js} A^s \right] - R_i{}^s \nabla_s A^i + R^j{}_s \nabla_j A^s \\
&= \nabla^j \left[R_{js} A^s \right],
\end{aligned}
\tag{2.81}
$$

where we again used the Lorenz gauge condition.

The covariant trace of Eq. (2.80) is not identically zero due to the presence of the Ricci tensor term in Eq. (2.81). On the other hand, this is precisely the extra term in Eq. (2.79). We can therefore conclude that Eq. (2.79) should be viewed as the correct Maxwell equation in curved spacetime since this is consistent with the charge conservation equation.

For matter fields describing particles with half integer spin like electrons, for instance, the situation is even more complicated. Depending on how a theory of gravity is constructed, one can arrive at two different theories. Both theories agree in the weak gravity limit and both predict almost the same physics. The only difference between these two theories is the treatment of spin and its coupling to gravity.

2.3. Weak Gravity

One of the main aims of the previous section was to motivate the correct form of the Einstein field equations as much as possible.[1] We will now do the reverse, namely check whether the Einstein field equations

---

[1] On a more philosophical note, one cannot derive any physical theory from scratch as ultimately experiments verify or falsify theories. We model nature as best as we can.

correctly reduce to the Newtonian equations when we assume slow motions and weak gravitational fields.

2.3.1. *Linearised Riemann and Ricci tensors*

We assume that the spacetime is described by a metric which is nearly the Minkowski metric, with small perturbation due to gravity, hence we write our metric as follows:

$$g_{ab} = \eta_{ab} + h_{ab}, \tag{2.82}$$

where η_{ab} is the Minkowski metric and h_{ab} are the small deviations from this flat space, they satisfy $|h_{ab}| \ll 1$.

The inverse metric is given by $g^{ab} = \eta^{ab} - h^{ab}$ which can be seen as follows:

$$\begin{aligned}
\delta_a^c = g_{ab}g^{bc} &= (\eta_{ab} + h_{ab})(\eta^{bc} - h^{bc}) \\
&= \delta_a^c + h_{ab}\eta^{bc} - h^{bc}\eta_{ab} + O(h^2) = \delta_a^c,
\end{aligned} \tag{2.83}$$

where we raise and lower the indices of h_{ab} using the Minkowski metric.

To first order in h_{ab} the Christoffel symbol is simply given by

$$\Gamma^n_{ab} = \frac{1}{2}\eta^{nd}\left(\partial_a h_{bd} + \partial_b h_{da} - \partial_d h_{ab}\right). \tag{2.84}$$

Next, we wish to compute the Riemann curvature tensor, given by Eq. (1.164). The terms containing the squares of the Christoffel symbols will be of second order and only the derivative terms will contribute to the lowest-order terms. A direct calculation gives

$$\begin{aligned}
R_{adb}{}^c &= \partial_d \Gamma^c_{ab} - \partial_a \Gamma^c_{db} \\
&= \frac{1}{2}\eta^{cs}\left(\partial_{bd}h_{sa} + \partial_{ad}h_{bs} - \partial_{sd}h_{ab} - \partial_{ba}h_{sd} - \partial_{da}h_{bs} + \partial_{sa}h_{db}\right) \\
&= \frac{1}{2}\eta^{cs}\left(\partial_{bd}h_{sa} - \partial_{ba}h_{sd} + \partial_{sa}h_{db} - \partial_{sd}h_{ab}\right). \tag{2.85}
\end{aligned}$$

For the Einstein field equations we need the Ricci tensor and the Ricci scalar which we are computing next. Summing over the second and fourth index in Eq. (2.85) gives

$$R_{ab} = R_{acb}{}^c = \frac{1}{2}\eta^{cs}\left(\partial_{bc}h_{sa} - \partial_{ba}h_{sc} + \partial_{sa}h_{cb} - \partial_{sc}h_{ab}\right)$$
$$= \frac{1}{2}\left(\partial_{bs}h^s{}_a + \partial_{as}h^s{}_b - \partial_{ba}h - \Box h_{ab}\right), \tag{2.86}$$

where $h = h^s{}_s$ is the trace of h_{ab}. We should pay particular attention to the last term which is $\eta^{cs}\partial_{sc}h_{ab} = \Box h_{ab}$. Here $\Box$ is the D'Alembertian or sometimes called wave operator. In Minkowski space we simply have

$$\Box\phi = \eta^{cd}\partial_{cd}\phi = \left(-\frac{\partial^2}{\partial t^2} + \Delta\right)\phi. \tag{2.87}$$

If ϕ is time-independent then $\Box\phi = \Delta\phi$ and we note that the Ricci tensor indeed contains the Laplace operator, as expected from Eq. (2.73).

We can introduce the trace reversed tensor $\bar{h}_{ab} = h_{ab} - \eta_{ab}h/2$ whose trace is given by $\bar{h} = h - 2h = -h$, as the name suggests. Using this tensor, the Ricci tensor (2.86) becomes

$$R_{ab} = \frac{1}{2}\left(\partial_{bs}\bar{h}^s{}_a + \partial_{as}\bar{h}^s{}_b + \frac{1}{2}\eta_{ab}\Box\bar{h} - \Box\bar{h}_{ab}\right). \tag{2.88}$$

Consequently, the Ricci scalar is given by

$$R = \partial_{bs}\bar{h}^{sb} + \frac{1}{2}\Box\bar{h}, \tag{2.89}$$

so that the Einstein tensor, in lowest order in h_{ab}, is given by

$$G_{ab} = R_{ab} - \frac{1}{2}\eta_{ab}R$$
$$= \frac{1}{2}\left(\partial_{bs}\bar{h}^s{}_a + \partial_{as}\bar{h}^s{}_b + \frac{1}{2}\eta_{ab}\Box\bar{h} - \Box\bar{h}_{ab} - \eta_{ab}\partial_{st}\bar{h}^{st} - \frac{1}{2}\eta_{ab}\Box\bar{h}\right)$$
$$= \frac{1}{2}\left(\partial_{bs}\bar{h}^s{}_a + \partial_{as}\bar{h}^s{}_b - \Box\bar{h}_{ab} - \eta_{ab}\partial_{st}\bar{h}^{st}\right). \tag{2.90}$$

This Einstein tensor (2.90) would simplify considerably if we could make terms of the form $\partial_s \bar{h}^s{}_i$ disappear. This looks very similar to the Lorenz gauge discussed previously in electromagnetism, so we need to investigate whether a particular coordinate system exists for which these terms do indeed vanish.

2.3.2.   *Gauge transformations*

Let us start with a small coordinate transformation

$$X^i = X'^i - \xi^i(X'^k), \tag{2.91}$$

where the vector ξ^i is assumed to be small at all points. Then

$$dX^i = dX'^i - \frac{\partial \xi^i}{\partial X'^k} dX'^k, \tag{2.92}$$

and hence we have

$$dX^i = dX'^i - \frac{\partial \xi^i}{\partial X'^k} dX'^k,$$

$$\frac{\partial X^i}{\partial X'^j} = \delta^i_j - \frac{\partial \xi^i}{\partial X'^j}. \tag{2.93}$$

Applying the tensor transformation rule (1.34) to the metric tensor $g_{ab} = \eta_{ab} + h_{ab}$ we find that

$$g'_{ab} = \eta_{ab} + h_{ab} - \partial_a \xi_b - \partial_b \xi_a = \eta_{ab} + h'_{ab}, \tag{2.94}$$

so that

$$h'_{ab} = h_{ab} - \partial_a \xi_b - \partial_b \xi_a. \tag{2.95}$$

We assumed that the ξ^i are small and therefore h'_{ab} is also small. The vector ξ^i has 4 degrees of freedom and therefore we will be able to fix 4 values of h'_{ab}. Let us raise both indices in Eq. (2.95) and also differentiate with respect to X'^a. We arrive at

$$\partial_a h'^{ab} = \partial_a h^{ab} - \Box \xi^b - \partial_a \partial^b \xi^a. \tag{2.96}$$

Using the trace reversed tensor, we have $\bar{h}^{ab} = h^{ab} - \eta^{ab} \bar{h}/2$ and hence Eq. (2.96) becomes

$$\partial_a \bar{h}'^{ab} = \partial_a \bar{h}^{ab} - \Box \xi^b. \tag{2.97}$$

Therefore, we can find a coordinate system where $\partial_a \bar{h}'^{ab} = 0$ by choosing $\Box \xi^b = \partial_a \bar{h}^{ab}$. This means we have an equivalent of the Lorenz gauge in General Relativity.

2.3.3.   Linearised Einstein field equations

Finally, we are able to state the Einstein field equations linear in the metric perturbation. Working in the equivalent of the Lorenz gauge we find

$$G_{ab} = -\frac{1}{2}\Box \bar{h}_{ab} = \kappa T_{ab}. \tag{2.98}$$

Our aim is to recover the Newtonian equations of gravity, and the simplest way to achieve this is to assume a static gravitational field where all quantities are time independent. We should also note that the linearised Einstein field equations are the starting point to the study of gravitational waves, another important prediction of General Relativity, which we briefly introduce in Sec. 2.3.4.

For the matter tensor we assume a perfect fluid given by Eq. (2.33). For non-relativistic matter $\rho c^2 \gg p$ and so the only non-vanishing component of the energy–momentum tensor is $T_{00} = \rho c^2$. Therefore, the linearised Einstein field equations (2.98) reduce to the single equation

$$\Box \bar{h}_{00} = -2\kappa \rho c^2. \tag{2.99}$$

For static fields $\Box \bar{h}_{00} = \Delta \bar{h}_{00}$, with $\Delta = \partial_{xx} + \partial_{yy} + \partial_{zz}$ being the Laplacian. Consequently, we arrive at an equation similar to the Newtonian one $\Delta \phi_N = 4\pi G\rho$, which reads

$$\Delta \bar{h}_{00} = -2\kappa \rho c^2. \tag{2.100}$$

Taking into account the Newtonian equation (2.8) we are led to make the identification

$$\bar{h}_{00} = -\frac{\kappa c^2}{2\pi G}\phi_N. \tag{2.101}$$

Reverting back to the unbarred quantity, we have

$$h_{00} = -\frac{\kappa c^2}{4\pi G}\phi_N, \quad h_{11} = h_{22} = h_{33} = -\frac{\kappa c^2}{4\pi G}\phi_N. \qquad (2.102)$$

and therefore, the solution of the linearised field equations yields the metric

$$ds^2 = -\left(1 + \frac{\kappa c^2}{4\pi G}\phi_N\right) + \left(1 - \frac{\kappa c^2}{4\pi G}\phi_N\right)\left[dx^2 + dy^2 + dz^2\right]. \qquad (2.103)$$

Comparison of this line element with Eqs. (2.59) and (2.66) implies that we must have

$$\frac{\kappa c^2}{4\pi G}\phi_N = 2\phi. \qquad (2.104)$$

This is a crucial relation. It fixes the coupling constant κ such that the theory reproduces correctly the Newtonian limit. Recall Eq. (2.66) which states $\phi_N = \phi c^2$ and implies the following choice for a consistent theory

$$\kappa = \frac{8\pi G}{c^4}, \qquad (2.105)$$

which is the main result of this calculation. In summary, we can now state the Einstein field equations again, with correct coupling constant

$$G_{ij} = \frac{8\pi G}{c^4}T_{ij}. \qquad (2.106)$$

For most practical purposes it is convenient to set $G = c = 1$, a convention used consistently in theoretical physics. The main reason is that it can be quite tricky to keep track of all the correct factors of G and in particular c in long calculations and it is often easier to reinsert these quantities at the end. In what follows we will stick to this convention and only re-introduce physical constants where needed.

2.3.4. *Gravitational waves*

We begin with the linearised Einstein field equations (2.98) in vacuum. To simplify the following discussion, we will consider $\bar{h}_{ab}$ depending on time t and the z-direction only. Under this assumption the linearised field equations become

$$\left(-\frac{\partial^2}{\partial t^2} + \frac{\partial^2}{\partial z^2}\right)\bar{h}_{ab} = 0, \tag{2.107}$$

which is a wave equation for the components $\bar{h}_{ab}$ and corresponds to plane waves travelling along the z-direction. Similar wave equations describe the propagation of electromagnetic radiation or the propagation of elastic waves. We immediately note that the wave speed equals the speed of light, so we can conclude that these gravitational waves travel at the speed of light.

In order to determine the solution of this wave equation, we use the ansatz

$$\bar{h}_{ab} = A_{ab}\cos(\omega z - \omega t), \tag{2.108}$$

where A_{ab} is some tensorial amplitude. This corresponds to waves travelling in the positive z-direction. One could also work with a plus sign and consider waves travelling in the negative z-direction. One can check directly that Eq. (2.108) satisfies the linearised Einstein field equations.

Our wave ansatz Eq. (2.108) contains the amplitude A_{ab} which is a rank-2 symmetric tensor. In general, this has 10 independent components since we work in four dimensions. Recall that we assume $\bar{h}_{ab}$ to satisfy the Lorenz gauge $\partial_a \bar{h}^{ab} = 0$ which will restrict the form of A_{ab}. Let us briefly consider the more general plane wave ansatz

$$\bar{h}^{ab} = A^{ab}\cos(k_i x^i), \tag{2.109}$$

where $x^i = (t, x, y, z)$ and we introduced the wave (co)vector k_i. Then, the gauge condition becomes

$$\partial_a \bar{h}^{ab} = \partial_a \left(A^{ab} \cos(k_i x^i) \right) = A^{ab} k_a = 0. \tag{2.110}$$

These are four equations in general, and hence A^{ab} can have at most six independent components. However, we can still make infinitesimal coordinate transformations of the form (2.91) with $\Box \xi^a = 0$ which leave the Lorenz gauge unchanged, see Eq. (2.97). Since ξ^a is a vector with 4 degrees of freedom, we can eliminate another four components from A^{ab}. Therefore, A^{ab} has two independent components which contain the physics of gravitational waves. In Eq. (2.108) we chose $k_i = (-\omega, 0, 0, \omega)$.

One can choose the ξ^a such that the gravitational wave solution travelling in the positive z-direction is given by

$$\bar{h}^{ab} = A^{ab} \cos(\omega z - \omega t) = \begin{pmatrix} 0 & 0 & 0 & 0 \\ 0 & A_+ & A_\times & 0 \\ 0 & A_\times & -A_+ & 0 \\ 0 & 0 & 0 & 0 \end{pmatrix} \cos(\omega z - \omega t). \tag{2.111}$$

This particular gauge is called the transverse-traceless gauge because the wave travels along the z-direction and distorts objects in the x and y directions, the transverse directions. Traceless refers to the fact that A^{ab} has zero trace.

The behaviour of freely falling nearby test particles under the influence of the gravitational wave is determined by the geodesic deviation equation. In Eq. (1.213) the tangent vector T^i now corresponds to the test particle's 4-velocity which we can choose to be $T^i = (1, 0, 0, 0)$. Therefore, only the components $R_{i00}{}^a$ of the Riemann curvature tensor enter the geodesic deviation equation, and

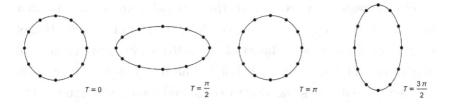

Fig. 2.3　Visualisation of the gravitational wave with polarisation A_+.

one finds in linear approximation

$$R_{100}{}^1 = -R_{200}{}^2 = -\frac{1}{2}A_+\omega^2\cos(\omega z - \omega t), \qquad (2.112)$$

$$R_{100}{}^2 = R_{200}{}^1 = -\frac{1}{2}A_\times\omega^2\cos(\omega z - \omega t). \qquad (2.113)$$

The other components vanish. Therefore, the vector N^i connecting nearby particles must have the form $N^i = (0, N^1, N^2, 0)$ which means that this vector is contained in the xy-plane and this plane is perpendicular to the propagation direction of the gravitational wave. The geodesics deviation equations become

$$\frac{D^2N^1}{D\lambda^2} = -\frac{1}{2}\left(A_+N^1 + A_\times N^2\right)\omega^2\cos(\omega z - \omega t), \qquad (2.114)$$

$$\frac{D^2N^2}{D\lambda^2} = -\frac{1}{2}\left(A_\times N^1 - A_+N^2\right)\omega^2\cos(\omega z - \omega t). \qquad (2.115)$$

The solutions to these equations describe harmonic oscillations of some vector $(N^i)_{\text{initial}}$ in the xy-plane. If we imagine a ring of test particles initially at rest, then the gravitational wave passing through the plane will distort this ring of particles to an elliptical shape. The lengths of the two semi-axes changes in time while the area enclosed by the particles remains constant, this is visualised in Fig. 2.3.

We should also note that the waves with $A_+ = 0$ and those with $A_\times = 0$ corresponds to two different polarisations of the waves, these two states are related by a $\pi/4$ or $45°$ rotation. This is in contrast to electromagnetic waves where the two polarisations are rotated by 90°. A very recent observation confirms the first direct detection of gravitational waves, see Abbott *et al.* (2016), emitted by the inspiral and merger of a pair of black holes. The first indirect observation of gravitational waves was due to Hulse and Taylor who studied the orbital decay of a binary pulsar system which decays energy in the form of gravitational waves.

2.4.   Variational Approach to General Relativity

It turns out that physical theories can generally be formulated using a variational approach. This means that one states the action or Lagrangian and derives the equations governing the theory. These equations are the Euler–Lagrange equations coming from the action, we briefly discussed this framework in Sec. 1.4. In classical mechanics the interpretation of the Lagrangian is very clear as it is the difference between kinetic and potential energy. In General Relativity, on the other hand, it is difficult to define the concept of kinetic and potential energy and so it is not clear which form the Lagrangian should take so that variational calculus will result in the Einstein field equations.

In 1915, Hilbert noted that the action

$$S_{\text{EH}} = \frac{1}{2\kappa} \int R\sqrt{-g}\,d^4x, \qquad (2.116)$$

yields the Einstein field equations when variations with respect to the metric tensor g_{ij} are considered. For this reason action (2.116) is called the Einstein–Hilbert action. The term $\sqrt{-g}\,d^4x$ is nothing but the proper 4-dimensional volume on the manifold, see Eq. (1.57). Recall that g stands for the determinant of the metric tensor, and $\kappa = 8\pi G/c^4$. The presence of κ can be explained by checking the dimensions of the quantities involved. The Ricci scalar has dimensions of inverse length squared, $1/\text{m}^2$, $g = \det(g_{ij})$ is dimensionless and d^4x has units of volume multiplied by time $\text{m}^3 \times \text{s}$. Therefore, the quantity $R\sqrt{-g}\,d^4x$ has dimensions $\text{m} \times \text{s}$. The coupling constant κ has units of $\text{s}^2/\text{kg}/\text{m}$ which means that the Einstein–Hilbert action has the correct dimensions of energy times time.

One way to motivate the Ricci scalar R in this action is that it is the simplest curvature scalar which can be constructed from the Riemann curvature tensor. This single observation immediately allows us to propose different theories which could be based on the square of the Ricci tensor or the squared Riemann tensor.

We recall the equation for the derivative of the determinant of the metric (1.99). Since variations and partial derivatives are closely related, we can write

$$\delta g = g\, g^{ab} \delta g_{ab}, \tag{2.117}$$

where we took into account that the metric is symmetric, and hence we arrive at

$$\delta\sqrt{-g} = \frac{1}{2}\sqrt{-g}\, g^{ab} \delta g_{ab} = -\frac{1}{2}\sqrt{-g}\, g_{ab} \delta g^{ab}. \tag{2.118}$$

Next, we need to compute the variation of the Ricci scalar with respect to the metric. We begin by noting that the Ricci scalar is given by $g^{ab}R_{ab}$ and that R_{ab} can be expressed solely in terms of the Christoffel symbols. Let us make a small change in the Christoffel symbol, a direct calculation using Eqs. (1.188) and (1.164) shows that we can write

$$\delta R_{ab} = \partial_d \delta\Gamma^d_{ab} - \partial_a \delta\Gamma^d_{db} + \delta\Gamma^d_{dc}\Gamma^c_{ab} + \Gamma^d_{dc}\delta\Gamma^c_{ab}$$
$$- \delta\Gamma^d_{ac}\Gamma^c_{db} - \Gamma^d_{ac}\delta\Gamma^c_{db}. \tag{2.119}$$

While the Christoffel symbol is not a tensor, the object $\delta\Gamma$ is in fact a tensor as it is the difference between two connections. We used this argument before when discussing the torsion tensor at the end of Sec. 1.3.1. Inspection of the right-hand side of (2.119) shows that it can be written as the difference of two covariant derivatives

$$\delta R_{ab} = \nabla_d \delta \Gamma^d_{ab} - \nabla_a \delta \Gamma^d_{db}. \tag{2.120}$$

Putting these calculations together, we can begin calculating the variations of the Einstein–Hilbert action with respect to the metric

$$\delta S_{\text{EH}} = \frac{1}{2\kappa} \int \left[\delta g^{ab} R_{ab} \sqrt{-g} + g^{ab} \delta R_{ab} \sqrt{-g} - \frac{1}{2} R g_{ab} \delta g^{ab} \sqrt{-g} \right] d^4 x$$

$$= \frac{1}{2\kappa} \int \left[R_{ab} - \frac{1}{2} R g_{ab} \right] \delta g^{ab} \sqrt{-g} \, d^4 x$$

$$+ \frac{1}{2\kappa} \int \left[g^{ab} \delta R_{ab} \sqrt{-g} \right] d^4 x, \tag{2.121}$$

where we can already recognise the left-hand side (or Einstein tensor) of the Einstein field equations. In the following we will show that the remaining term is in fact a total derivative term which will not contribute to the field equations, analogously to Sec. 1.4 we keep the end points fixed. Using Eq. (2.120), the last term of (2.121) is

$$\frac{1}{2\kappa} \int \left[\sqrt{-g} \, \nabla_d (g^{ab} \delta \Gamma^d_{ab}) - \sqrt{-g} \, \nabla_a (g^{ab} \delta \Gamma^d_{db}) \right] d^4 x. \tag{2.122}$$

In Exercise 1.12, we showed the identity $\sqrt{-g} \, \nabla_i A^i = \partial_i (\sqrt{-g} \, A^i)$ which we can now apply to both terms in the integrand and arrive at

$$\frac{1}{2\kappa} \int \left[\partial_d (\sqrt{-g} \, g^{ab} \delta \Gamma^d_{ab}) - \partial_a (\sqrt{-g} \, g^{ab} \delta \Gamma^d_{db}) \right] d^4 x. \tag{2.123}$$

Now, we can clearly see that both terms are total derivatives and therefore these will not contribute to the equations of motion. Therefore, we have the desired result

$$\delta S_{\text{EH}} = \frac{1}{2\kappa} \int \left[R_{ab} - \frac{1}{2} R g_{ab} \right] \delta g^{ab} \sqrt{-g} \, d^4 x. \tag{2.124}$$

In order to derive the right-hand side of the Einstein field equations, we need to add an action describing the matter. Formally we write

$$S_{\text{matter}} = \int \mathcal{L}_{\text{matter}} \, d^4x, \qquad (2.125)$$

which means we can write the variations as

$$\delta S_{\text{matter}} = \int \delta \mathcal{L}_{\text{matter}}(\psi, \partial_i \psi) \, d^4x, \qquad (2.126)$$

where we assume that the matter Lagrangian depends on some fields ψ and their first partial derivatives. We define the energy–momentum tensor by

$$T_{ab} := -\frac{2}{\sqrt{-g}} \frac{\delta \mathcal{L}_{\text{matter}}}{\delta g^{ab}}, \quad \text{or} \quad \delta \mathcal{L}_{\text{matter}} = -\frac{1}{2} T_{ab} \sqrt{-g} \, \delta g^{ab}, \quad (2.127)$$

so that we can now formulate the complete field equations using the variational approach. We consider the total action $S_{\text{total}} = S_{\text{EH}} + S_{\text{matter}}$, its variations with respect to the metric are given by

$$\delta S_{\text{total}} = \int \left[\frac{1}{2\kappa} \left(R_{ab} - \frac{1}{2} R g_{ab} \right) - \frac{1}{2} T_{ab} \right] \delta g^{ab} \sqrt{-g} \, d^4x. \quad (2.128)$$

Requiring $\delta S_{\text{total}} = 0$ yields the Einstein field equations

$$R_{ab} - \frac{1}{2} R g_{ab} = \kappa T_{ab}. \qquad (2.129)$$

Additionally, we can consider the variation of S_{total} with respect to the matter fields, as in Eq. (1.229), which gives the equations of motion of the matter. Recall that these equations are not independent from the Einstein equations since the twice contracted Bianchi identities imply $\nabla_a T^{ab} = 0$.

2.5. Further Reading

General Relativity

Early work in General Relativity focused on finding exact solutions of physical importance to the Einstein field equations. This research moved on towards a systematic study of exact solutions taking into account various symmetry properties which can be imposed on the metric which in turn simplify the field equations. This has been a prosperous activity for many decades that led to the discovery of a vast number of exact solutions, many of which are collected in Stephani *et al.* (2003). The authors of this book state that they collected a total of over 6,000 references (up to 1999) which dealt with exact solutions. New exact solutions of the Einstein field equations, with and without matter sources, are still being discovered.

Substantial progress in General Relativity emerged from a detailed treatment of the mathematical aspects of the field equations. In particular the so-called initial value formulation was an important achievement in this direction. The interested reader should refer to Part II of the book by Wald (1984) which discusses various advanced topics of General Relativity which are beyond an introductory course. Also highly recommended are the books by Weinberg (1972) and Hawking and Ellis (1973), and the very comprehensive text by Misner *et al.* (1973), its 1,200 pages can be quite overwhelming though! Gravitational waves are discussed in detail in the book by Maggiore (2007). A more recent book which emphasises the many exciting mathematical aspects of General Relativity is by Choquet-Bruhat (2008). The discovery of General Relativity was now over 100 years ago, a centennial perspective of this subject discussing the many current research activities is by Ashtekar *et al.* (2015).

Another exciting aspect of General Relativity since its formulation are the many attempts to modify and extend the original theory. Einstein himself actively contributed to this field, and so did many others. In the subsequent text there will be some brief discussions about some of the various extensions and modifications of Einstein's original theory. The principal aim is to highlight the conceptual framework that was established to formulate General Relativity which itself motivates further studies beyond the actual theory.

More geometry — torsion and non-metricity. We start by recalling the definitions of torsion and non-metricity. We can define torsion to be the skew-symmetric part of the connection. We have $\nabla_a \nabla_b f = \partial_a \partial_b f - \Gamma_{ab}{}^i \partial_i f$ so that

$$2T_{ba}{}^i \partial_i f = \left(\Gamma_{ba}{}^i - \Gamma_{ab}{}^i\right) \partial_i f, \tag{2.130}$$

and therefore we can simply write

$$T_{ab}{}^i = \frac{1}{2}\left(\Gamma_{ab}{}^i - \Gamma_{ba}{}^i\right). \tag{2.131}$$

The object of non-metricity is defined by $Q_{aij} = \nabla_a g_{ij}$. As in Theorem 1.1 we can find the connection and express it in terms of the Christoffel symbol, the torsion tensor and the non-metricity. A direct calculation gives

$$\Gamma^n_{ij} = \left\{{}^n_{ij}\right\} + T_{ij}{}^n - T_j{}^n{}_i + T^n{}_{ij} + \frac{1}{2}\left(Q^n{}_{ij} - Q_{ij}{}^n - Q_{ji}{}^n\right). \tag{2.132}$$

This general connection is the basis for so-called metric affine gauge theories of gravity which have been researched in great detail. The use of gauge symmetries in gravity is natural in the sense that the other three fundamental forces are formulated as gauge field theories. A good introductory textbook into this subject is by Blagojevic (2001) which also covers the basic ideas of other modifications. Those who wish to read selected original papers with commentaries, putting those papers into the context of the entire research field, are probably well advised by Blagojevic and Hehl (2012). In this context, it is also worth pointing out that there exits an equivalent formulation of General Relativity based on the torsion tensor rather than the metric. The interested reader is referred to the article by Maluf (2013) and the book by Aldrovandi and Pereira (2013).

More dimensions — Kaluza–Klein and beyond. Instead of considering spaces with torsion and non-metricity, one could also consider a manifold with metric compatible connection without torsion but with more than four spacetime dimensions. This first such attempt goes back to Kaluza and Klein in the 1920s, see again Blagojevic (2001, Chap. 10). The basic idea is quite simple. Let us consider a 5D manifold, then the metric will have 15 independent components, five more than the 4D metric. One can now conjecture that the electromagnetic 4-vector potential A_i and an additional scalar field are the new degrees of freedom introduced in this theory. Since we only observe three spatial dimensions and one time dimension, one has to introduce an additional ingredient to reconcile this with a 5D theory. One generally assumes this extra dimension to be very small. However, there is no observational evidence for the existence of extra dimensions. Thoughts along these lines have also motivated additional research by further increasing the number of dimensions. For instance, String Theory is formulated in a 26D space which reduces to 10 dimensions when super-symmetry is taken into account, a good introductory textbook is by Zwiebach (2009). Super-symmetry, super-gravity and String Theory are also discussed in Blagojevic (2001, Chaps. 9 and 11).

More curvature — higher-order theories. When discussing the Einstein–Hilbert action in the variational approach to General Relativity, we noted that the Ricci scalar R is the simplest possible curvature scalar to be used in a Lagrangian. This motivates the study of theories which contain higher order curvature terms in the Lagrangian. One model which has received considerable attention during the previous decade is based on the idea of replacing the Ricci scalar by an arbitrary function of the Ricci scalar in the Einstein–Hilbert action. This function is normally denoted by $f(R)$ and therefore one speaks of $f(R)$-gravity, see Felice and Tsujikawa (2010); Sotiriou and Faraoni (2010) for two reviews on the subject.

The recommended literature is by no means complete, it simply reflects the author's suggestions for further reading.

2.6. Exercises

Some physics background

Exercise 2.1. Derive the pressure distribution inside a Newtonian star with constant density. Can this star be arbitrarily compact? Use M/R as a measure of compactness.

Exercise 2.2. Show that Eq. (2.27) is equivalent to Eq. (2.23).

Exercise 2.3 (annoying). The positioning of the factors of c in Sec. 2.1.4 is quite subtle and depends on various choices made. The crucial one being that we worked with coordinates $X^i = (ct, x, y, z)$ so that our four coordinates have the same units of length, then the Minkowski metric is $\eta_{ij} = \mathrm{diag}(-1, 1, 1, 1)$.

Let us now insist on coordinates $X^i = (t, x, y, z)$ and $\eta_{ij} = \mathrm{diag}(-1, 1, 1, 1)$. How does Eq. (2.34) have to be re-written and what is u^i in the rest frame and in general?

Exercise 2.4. Consider Lorentz boosts along the x-direction. The Lorentz transformation can be thought of as hyperbolic rotations in the sense that the transformation can be written as

$$ct' = ct\cosh(\zeta) - x\sinh(\zeta), \tag{2.133}$$

$$x' = x\cosh(\zeta) - ct\sinh(\zeta), \tag{2.134}$$

with $y' = y$ and $z' = z$. Find the matrix form of $L^a{}_b$ and show that $\det(L^a{}_b) = 1$. Next, show that the Minkowski metric is indeed invariant under this transformation. Find the relationships between β, γ and v, and the parameter ζ which is called rapidity. Finally, determine the angle α from Fig. 2.1 in terms of the rapidity.

Exercise 2.5. Follow on from Exercise 2.4. For simplicity we will neglect the y and z directions and work in the t, x plane only. Instead of working with the time coordinate ct we will now use the imaginary time coordinate ict. Let $R(\theta)$ be a 2D rotation matrix, show that

$$R(i\theta) \begin{pmatrix} ict \\ x \end{pmatrix}, \tag{2.135}$$

corresponds to a Lorentz transformation that is compatible with Eqs. (2.133) and (2.134).

Exercise 2.6 (takes time). Consider two Lorentz boosts along the x-direction with respective velocities v_1 and v_2, and rapidities ζ_1 and ζ_2. Show that the rapidity of the overall boost ζ (with velocity v) is simply their sum $\zeta = \zeta_1 + \zeta_2$. Use this to derive the velocity addition formula

$$v = \frac{v_1 + v_2}{1 + v_1 v_2/c^2}. \tag{2.136}$$

Exercise 2.7. Compute the components of the electromagnetic energy–momentum tensor (2.50) and express the result in terms of the electric field $\boldsymbol{E}$ and the magnetic field $\boldsymbol{B}$. Which components of T^{ij} are related to the Poynting vector $\boldsymbol{S} = \frac{c}{4\pi}(\boldsymbol{E} \times \boldsymbol{B})$? In Physics, the Poynting vector represents the directional energy flux density (the rate of energy transfer per unit area) of an electromagnetic field.

Exercise 2.8. The action of electromagnetism with source term is given by

$$S = \int \left(\frac{1}{\alpha} F^{ab} F_{ab} + A_b J^b \right) \sqrt{-g}\, d^4 x, \tag{2.137}$$

with $F_{ab} = \partial_a A_b - \partial_b A_a$. The geometrical setting is Minkowski space. Derive the Maxwell equations (2.28) by computing variations with respect to A_i and determine α.

Geometry and gravity

Exercise 2.9. Show that the vacuum ($T_{ij} = 0$) Einstein field equations imply that the Ricci scalar vanishes.

Exercise 2.10. Show that Eq. (2.77) implies the energy–momentum conservation equation $\nabla^i T_{ij} = 0$.

Exercise 2.11. Find the equivalent to Eq. (2.76) in the presence of the cosmological constant.

Exercise 2.12. Let $\mathcal{F}^{ij}$ be a skew-symmetric tensor $\mathcal{F}^{ij} = -\mathcal{F}^{ji}$. Show that the equation $\nabla_i \mathcal{F}^{ij} = 0$ implies a conservation law. This means it can be written as $\partial_i \mathcal{J}^i = 0$ for a suitably chosen $\mathcal{J}^i$.

Exercise 2.13. The quantity $2GM/(c^2 r)$ in Eq. (2.58) is dimensionless. Let α be some number of order one, the combination $\alpha \Lambda r^2$ is also dimensionless. Suggest the form of the metric with cosmological term and argue for the form of Newton's force law with Λ. Determine a simple upper bound on Λ using the Solar System.

Weak gravity

Exercise 2.14. Show that the definition for the trace reversed tensor $\bar{h}_{ab} = h_{ab} - \eta_{ab} h/2$ implies the relation $h_{ab} = \bar{h}_{ab} - \eta_{ab} \bar{h}/2$.

Exercise 2.15. Derive the transformation property (2.97) for $\bar{h}^{ab}$ under small coordinate transformations directly from the transformation property of h^{ab} beginning with Eq. (2.91).

Variational Approach to General Relativity

Exercise 2.16. Before Eq. (2.117) we wrote 'variations and partial derivatives are closely related'. Consider a sufficiently smooth function $f(x)$ depending on the independent variables x. Write the first few terms of the Taylor series of $f(x + \delta x)$, assuming that $\delta x \ll 1$. Think of δx as the small quantity h used in the definition of the derivative. Now introduce the notation δf for the difference between the changed and unchanged quantity, $\delta f = f(x + \delta x) - f(x)$ and rewrite the expression for the Taylor series. Establish the relationship between $\delta f / \delta x$ and the derivative df/dx.

Exercise 2.17 (hard). Starting with

$$S = \int \frac{1}{16\pi} F^{ab} F_{ab} \sqrt{-g} \, d^4 x, \qquad (2.138)$$

derive the energy–momentum tensor of the electromagnetic field in General Relativity using the calculus of variations.

Exercise 2.18. Consider a gravitational action that depends not just on the Ricci scalar but also the scalar $R_{ij} R^{ij}$. Argue that actions of this type yield theories that contain derivatives of higher than second order.

3

Schwarzschild Solutions

In this chapter, we are discussing the famous Schwarzschild solutions. It was discovered shortly after the formulation of General Relativity and was the first non-trivial solution of the Einstein field equations. The key idea was to assume a static and spherically symmetric space and study the field equations using these assumptions. It turns out that the field equations simplify considerably, and we will be able to find explicit solutions with and without matter.

3.1. Spherical Symmetry and Birkhoff's Theorem

Any symmetry of a manifold can be rigorously defined in differential geometry, however, it is possible to argue for the correct form of the metric differently. Let us begin with the standard spherical polar coordinates and let us compute the $ds^2 = dx^2 + dy^2 + dz^2$ using these coordinates. A direct calculation (see Example 1.4 in Sec. 1.2.3) gave

$$ds^2 = dr^2 + r^2 d\theta^2 + r^2 \sin^2\theta d\phi^2. \tag{3.1}$$

We can now argue that any spherically symmetric metric in General Relativity must contain the line element of the surface of the sphere. One often writes $d\Omega^2 = d\theta^2 + \sin^2\theta d\phi^2$ for this line element of the sphere. Hence, choosing coordinates $X^i = \{t, r, \theta, \phi\}$, we should be able to write any such metric in the form

$$ds^2 = -e^{A(t,r)}dt^2 + e^{B(t,r)}dr^2 + r^2 d\Omega^2. \tag{3.2}$$

The functions $A(t,r)$ and $B(t,r)$ are determined by solving the Einstein field equations and they depend on the choice of matter. The use of the exponential function is purely for convenience, it simplifies the subsequent field equations and makes them more manageable.

Let us now consider the exterior gravitational field of a spherically symmetric star for instance. Outside the star, we have a vacuum and therefore $T_{ab} = 0$, and the Einstein field equations reduce to $G_{ab} = 0$. Computing the Einstein tensor components explicitly is rather involved. First, one uses the metric (3.2) to find all the non-vanishing Christoffel symbol components. These can then be used to compute the Riemann tensor components which in turn lead to the Ricci tensor components. Finally, one can calculate the components of the Einstein tensor.

First, we consider the off-diagonal equation $G_{01} = 0$ which is given by

$$G_{01} = \frac{1}{r}\dot{B} = 0, \tag{3.3}$$

where the dot means differentiation with respect to time t. Therefore, the function B must be independent of time and hence a function of r only. Next we consider the combination $\exp(-A+B)G_{00}+G_{11} = 0$ which gives

$$\frac{1}{r}A' + \frac{1}{r}B' = 0, \tag{3.4}$$

which implies that $A' = -B'$ where the prime means differentiation with respect to r. Since B is a function of r only, it follows that A must also be a function of r only. Therefore, both functions are independent of time and the vacuum spacetime is static. What we have just shown is part of what is known as Birkhoff's theorem which states: *Every spherically symmetric solution of the vacuum field equations is static and asymptotically flat. The unique exterior solution is the Schwarzschild solution.* (We will show the rest of this statement shortly.) This means that the assumption of spherical symmetry implies staticity in the vacuum region which has some important implications.

Let us consider a pulsating (spherically symmetric) star, by this we mean an object whose radius changes over time. We know that the exterior gravitational field of this star is static and therefore such a pulsating object cannot create any gravitational waves. Even the formation of a black hole would not generate any gravitational radiation provided that spherical symmetry is perfectly maintained. This also tells us that a good astrophysical candidate for the emission of gravitational radiation would be a binary system, two objects orbiting rapidly around their centre of mass.

3.2. The Schwarzschild Solution

We showed that the most general spherically symmetric metric in vacuum is of the form

$$ds^2 = -e^{A(r)}dt^2 + e^{B(r)}dr^2 + r^2 d\Omega^2, \tag{3.5}$$

where our coordinates are as before $X^i = \{t, r, \theta, \phi\}$. The Einstein tensor components for this metric are given by

$$G_{00} = \frac{1}{r^2}e^{A-B}(e^B + rB' - 1), \tag{3.6}$$

$$G_{11} = \frac{1}{r^2}(1 + rA' - e^B), \tag{3.7}$$

$$G_{22} = \frac{1}{2}e^{-B}r^2\left(A'' + (A' - B')\left(\frac{1}{r} - \frac{1}{2}A'\right)\right), \tag{3.8}$$

$$G_{33} = \sin^2\theta\, G_{22}. \tag{3.9}$$

Our aim is to solve the vacuum field equations $G_{ab} = 0$. It seems that Eqs. (3.6)–(3.9) are three equations for the two unknown functions A and B. However, recall the twice contracted Bianchi identity, Theorem 1.6, which states that the Einstein tensor has to satisfy additional equations. This implies that the equation $G_{22} = 0$ can in fact be derived from the other two field equations, and hence it suffices to consider Eqs. (3.6) and (3.7). The equations $G_{ab} = 0$ are now equivalent to

$$e^B + rB' - 1 = 0, \tag{3.10}$$

$$-e^B + rA' + 1 = 0, \tag{3.11}$$

and it is not difficult to solve these equations.

We start by noting that Eq. (3.10) can be written as follows:

$$e^B + rB' - 1 = e^B \frac{d}{dr}(r - re^{-B}), \tag{3.12}$$

and therefore Eq. (3.10) can be integrated to give

$$r - re^{-B} = C \quad \Rightarrow \quad e^{-B} = 1 - \frac{C}{r}, \tag{3.13}$$

where C is a constant of integration. Addition of Eqs. (3.10) and (3.11) gives $r(A' + B') = 0$ from which we conclude $A = -B$. Note that there is a second constant of integration, however, this can always be set to one by rescaling the time coordinate. Therefore, we arrive at

$$e^A = e^{-B} = 1 - \frac{C}{r}, \tag{3.14}$$

which implies that the metric is of the form

$$ds^2 = -\left(1 - \frac{C}{r}\right) dt^2 + \left(1 - \frac{C}{r}\right)^{-1} dr^2 + r^2 d\Omega^2. \tag{3.15}$$

This metric depends on the constant C for which we need to find a physical interpretation. Recall that the line element Eq. (3.15) describes the exterior gravitational field of any spherically symmetric source. In Newtonian gravity this field is uniquely characterised by the mass M of the object and so we suspect C to be related to the mass.

One neat way of arriving at the correct interpretation of C is to re-do the above calculation with Newton's constant G and the speed of light c. In this case our metric Eq. (3.5) should be written in the form

$$ds^2 = -e^{A(r)/c^2} c^2 dt^2 + e^{B(r)} dr^2 + r^2 d\Omega^2, \qquad (3.16)$$

and the constant of integration C we would write as GC/c^4 so that this C would have dimensions of mass. As before we find

$$e^{A/c^2} = e^{-B} = 1 - \frac{GC}{c^4 r}. \qquad (3.17)$$

However, let us now compute the Christoffel symbol components for Eq. (3.16) and consider the limit $c \to \infty$ which corresponds to the Newtonian limit. It turns out there is only one non-vanishing component which depends on the constant C. One finds

$$\Gamma^1_{00} = \frac{(C/2)\,G}{r^2}, \qquad (3.18)$$

which is part of Newton's force law provided we choose $2M = C$, compare with Eq. (2.4). The other non-vanishing Christoffel symbol components are due to our choice of coordinates. Alternatively, we can recall Eqs. (2.59) and (2.66) which would result in the same conclusion. With this identification made, we can now state the famous Schwarzschild solution

$$ds^2 = -\left(1 - \frac{2M}{r}\right) dt^2 + \left(1 - \frac{2M}{r}\right)^{-1} dr^2 + r^2 d\Omega^2. \qquad (3.19)$$

One of the first observations is that this metric is singular at $r = 2M$ and at $r = 0$. At this point it is not clear whether these are coordinate singularities due to our bad choice of coordinates or whether these are true spacetime singularities where the curvature tensor components diverge. The vacuum field equations can be written as $R_{ij} = 0$, so the Ricci tensor vanishes identically. However, this does not imply the vanishing of curvature as expressed by the Riemann curvature tensor.

Recall that the Riemann tensor has 20 independent components in four dimensions while the Ricci tensor has only 10, the other 'half' of the Riemann tensor is encoded in the Weyl tensor, see Definition 1.22. In order to check for true spacetime singularities one has to study the entire Riemann curvature tensor.

The Schwarzschild solution was the first non-trivial exact solution of the Einstein field equations. It was found by Karl Schwarzschild in 1916 less than a year after the Einstein field equations were formulated. This solution has been extensively studied during the last century and it is fair to say that it is still regarded as one of the most important solutions of the theory. It can be used to derive predictions of General Relativity which can be tested in the Solar System and these predictions are different to those from Newtonian gravity. Observations are in excellent agreement with Einstein's theory. The Schwarzschild solution also is at the heart of black hole physics as it can be interpreted as the exterior of a non-rotating black hole. In the following we will look at many interesting aspects of the Schwarzschild solution.

The radius $r = 2M$ where the metric is singular is often called the Schwarzschild radius, re-introducing physical constants, we define the Schwarzschild radius by

$$r_S = \frac{2GM}{c^2}. \tag{3.20}$$

At this point it is useful to check the significance of this radius for an object like the Sun. Taking the solar mass to be $M_\odot = 1.99 \times 10^{30}$ kg, we find

$$(r_S)_\odot = 2.95 \, \text{km}. \tag{3.21}$$

The physical radius of the Sun is $R_\odot = 6.955 \times 10^5$ km. Hence, the Schwarzschild radius of the Sun is much smaller than the actual radius of the Sun. The Schwarzschild solution describes only the exterior gravitational field of the Sun starting from its boundary, so the Schwarzschild radius has no physical meaning in this context. The interior of the Sun is described by a solution to the field equations with matter, we discuss one such solution in the following section.

Let us briefly think about a very massive collapsing star. It is possible for the gravitational force to be stronger than any of the other forces which would prevent this collapse. In such a situation the collapse continues beyond the Schwarzschild radius and the $r_S = 2M$ surface becomes physically meaningful. When this happens the object forms what is known as a black hole. It is widely agreed that (supermassive) black holes should exist at the centres of galaxies. Note that the absence of any mechanism to stop the continued collapse implies that eventually the entire mass would be concentrated at the centre of the black hole.

Using spatial Cartesian coordinates, the Schwarzschild metric can be written in the form

$$ds^2 = -\frac{\left(1 - \frac{M}{2r}\right)^2}{\left(1 + \frac{M}{2r}\right)^2} dt^2 + \left(1 + \frac{M}{2r}\right)^4 \left(dx^2 + dy^2 + dz^2\right), \quad (3.22)$$

where $r^2 = x^2 + y^2 + z^2$ is the Euclidean distance from the origin.

3.3. The Schwarzschild Interior Solution

Next we are interested in solving the Einstein field equations in the presence of some matter. For simplicity we assume this to be a perfect fluid of constant density, this serves as a rough model of an incompressible star, for instance a neutron star. As before, we assume a static and spherically symmetric metric of the form Eq. (3.5). The energy–momentum tensor Eq. (2.34) in a non-flat space takes the form

$$T_{ij} = \rho\, u_i u_j + p(g_{ij} + u_i u_j). \quad (3.23)$$

For our given metric the fluid's 4-velocity is $u_0 = -e^{A(r)/2}$, $u_1 = u_2 = u_3 = 0$ so that the components of T_{ij} are given by $T_{00} = \rho e^A$, $T_{11} = p e^{B(r)}$, $T_{22} = p r^2$ and $T_{33} = \sin^2\theta\, T_{22}$. The resulting three independent Einstein field equations are

$$\frac{1}{r^2} e^{A-B}(e^B + rB' - 1) = 8\pi\rho\, e^A, \qquad (3.24)$$

$$\frac{1}{r^2}(1 + rA' - e^B) = 8\pi p\, e^B, \qquad (3.25)$$

$$\frac{1}{2} e^{-B} r^2 \left(A'' + (A' - B')\left(\frac{1}{r} - \frac{1}{2}A'\right) \right) = 8\pi p\, r^2. \qquad (3.26)$$

The final equation $G_{33} = 8\pi\, T_{33}$ differs from Eq. (3.26) only by a factor of $\sin^2\theta$.

Let us begin with Eq. (3.24), we divide by the common factor of e^A and multiply the entire equation by r^2. This yields

$$1 + rB'e^{-B} - e^{-B} = 8\pi r^2\rho. \qquad (3.27)$$

Following our observation that led to Eq. (3.12) we can rewrite the previous equation as

$$\frac{d}{dr}\left(r - re^{-B}\right) = 8\pi\rho r^2, \qquad (3.28)$$

which strongly suggests that we should integrate this relation. The right-hand side is identical to the Newtonian mass definition in spherical symmetry and hence we define

$$m(r) = \int_0^r 4\pi\rho(\bar{r})\bar{r}^2 d\bar{r}, \qquad (3.29)$$

as the mass up to radius r. If, moreover, the density is assumed to be constant $\rho(r) = \rho_0$, then the mass up to r simply becomes

$$m(r) = \frac{4\pi}{3}\rho_0 r^3. \qquad (3.30)$$

Using the Newtonian mass definition (3.29), we can solve Eq. (3.28) for one of the metric functions and find

$$e^{-B} = 1 - \frac{2m(r)}{r}, \tag{3.31}$$

which is in nice agreement with the Schwarzschild solution given by (3.19). Note that we set the constant of integration which appears to zero in order to avoid a singular solution for small r. Next, assuming a constant density gives the explicit form

$$e^{-B} = 1 - \frac{8\pi\rho_0}{3}r^2, \tag{3.32}$$

and we note that the spatial part of the metric describing an incompressible star is now uniquely determined. We are left with one function to be determined in the metric

$$ds^2 = -e^{A(r)}dt^2 + \frac{dr^2}{1 - (8\pi\rho_0/3)r^2} + r^2 d\Omega^2. \tag{3.33}$$

We will encounter a very similar spatial part of a metric again when discussing Cosmology in Chap. 4. We still need to find the pressure function and the other metric function to fully solve the field equations.

At this point it is best to state the energy–momentum conservation equation $\nabla_a T^a_b = 0$ which for our matter source leads to

$$p' + \frac{1}{2}A'(\rho + p) = 0. \tag{3.34}$$

Note that this equation is not independent of the field equations but a consequence of them due to the twice contracted Bianchi identities, Theorem 1.6. We will now eliminate the quantity A' from the conservation equation (3.34) and the second field equation (3.25). This gives

$$p' + \frac{e^B}{2r}(\rho + p)(8\pi p r^2 + 1 - e^{-B}) = 0. \tag{3.35}$$

Since we already determined the function e^B, this can be rewritten as

$$p' = -r\frac{(4\pi p + m/r^3)(\rho + p)}{1 - 2m/r},$$

(3.36)

which is the famous Tolman–Oppenheimer–Volkoff equation. Let us consider the non-relativistic limit of this equation by assuming $p \ll \rho$ and $2m/r \ll 1$ so that it becomes

$$p' = -\frac{m\rho}{r^2},$$

(3.37)

which is the structure equation of Newtonian astrophysics, see also Exercise 2.1.

When assuming a constant density distribution, the Tolman–Oppenheimer–Volkoff equation simplifies considerably and becomes a separable first-order ordinary differential equation which reads

$$p' = -r\frac{4\pi(p + \rho_0/3)(\rho + p)}{1 - 8\pi/3\rho_0 r^2}.$$

(3.38)

When solving this equation we choose the constant of integration such that $p(r = 0) = p_c$ because the central pressure of our hypothetical star is a physically useful quantity. The solution of Eq. (3.38) can be written in the following form:

$$p(r) = \rho_0 \frac{(3p_c + \rho_0)\sqrt{1 - \frac{8\pi\rho_0}{3}r^2} - (p_c + \rho_0)}{3(p_c + \rho_0) - (3p_c + \rho_0)\sqrt{1 - \frac{8\pi\rho_0}{3}r^2}}.$$

(3.39)

Alternatively, we can use the mass equation (3.30) to rewrite Eq. (3.39) as follows:

$$p(r) = \rho_0 \frac{(3p_c + \rho_0)\sqrt{1 - \frac{2m(r)}{r}} - (p_c + \rho_0)}{3(p_c + \rho_0) - (3p_c + \rho_0)\sqrt{1 - \frac{2m(r)}{r}}}.$$

(3.40)

The function $p(r)$ is decreasing and takes its maximum value at $r = 0$, the centre of the star. From a physical point of view this is very satisfying. The pressure decreases as one moves from the centre towards the surface of the star. We will define this surface or boundary of the star to be the vanishing pressure surface. This means, we define the radius R of the star by the relation $p(r = R) = 0$. We also denote the total mass of the star by M so that $M = m(R)$, capital letters are used for total quantities.

Using $p(r = R) = 0$ in Eq. (3.39) allows us to find an expression relating the total mass M, the radius R and the central pressure p_c and energy density ρ_0 which is given by

$$\sqrt{1 - \frac{2M}{R}} = \frac{(p_c + \rho_0)}{(3p_c + \rho_0)}. \tag{3.41}$$

This equation has one particularly neat implication which has no equivalent in Newtonian astrophysics. Let us solve this equation for the central pressure p_c, we find

$$p_c = \rho_0 \frac{1 - \sqrt{1 - \frac{2M}{R}}}{3\sqrt{1 - \frac{2M}{R}} - 1}, \tag{3.42}$$

which puts a constraint on the mass–radius ratio of our star provided we assume $p_c < \infty$. A condition of this form is physically sensible, we would like to describe the centre of an astrophysical object with some regular form of matter, in particular we wish to require that all physical quantities are finite. For the central pressure to be finite means that the denominator of Eq. (3.42) must be larger than zero which implies

$$\sqrt{1 - \frac{2M}{R}} > \frac{1}{3} \quad \Rightarrow \quad \frac{2M}{R} < \frac{8}{9}. \tag{3.43}$$

Therefore, in General Relativity we cannot arbitrarily increase the mass of an object while keeping the radius fixed. In other words there exists a bound on the compactness of an astrophysical object, this result goes back to Buchdahl. In turn this implies that an object of a given mass must have a minimal radius so that the inequality (3.43) is satisfied. Reinstating physical constants, we would write the Buchdahl inequality as

$$\frac{2GM}{c^2 R} < \frac{8}{9}.$$ (3.44)

Before continuing, let us state, for completeness, the metric of this interior solution. The metric function A can be found by integrating Eq. (3.34). For a constant density distribution, this integration yields

$$A(r) = \log\left(\frac{C}{\rho_0 + p(r)}\right)^2,$$ (3.45)

where C is a constant of integration. It can be chosen arbitrarily, and its value can be changed by rescaling the time coordinate used in the metric. A convenient choice is to take $A(r = 0) = 0$ so that $C = \rho_0 + p_c$. Inserting this solution into Eq. (3.33) gives the Schwarzschild interior metric

$$ds^2 = -\left(\frac{\rho_0 + p_c}{\rho_0 + p(r)}\right)^2 dt^2 + \frac{dr^2}{1 - (8\pi/3)\rho_0 r^2} + r^2 d\Omega^2,$$ (3.46)

where the function $p(r)$ is given by (3.39) or (3.40).

As in the previous section, let us evaluate this inequality for the Sun, we have

$$\frac{2GM_\odot}{c^2 R_\odot} = 4.25 \times 10^{-6},$$ (3.47)

which satisfies the Buchdahl inequality. Interestingly, despite the Buchdahl inequality being satisfied by about five orders of magnitude, very few stars are known whose masses exceed $200\ M_\odot$. One should keep in mind though that the Buchdahl inequality is valid only in the context of static and spherically symmetric objects which is violated by any realistic astrophysical object.

This has further implications in astrophysics. When observing spiral galaxies and investigating the motion of objects near their centres, one is required to place a very heavy, yet small object into their respective centres to account for that motion. It turns out that this hypothetical object grossly violates the inequality (3.42). In fact, simple estimates indicate that the physical radius of this object is smaller than the corresponding Schwarzschild radius. Therefore, no known matter type would be able to describe such an object and only a black hole is compatible with observations.

3.4. Geodesics in Schwarzschild Spacetime

The Schwarzschild solution describes the exterior gravitational field of a spherically symmetric body, we are now interested in understanding the motion of test particles in the Schwarzschild spacetime. In practical terms, we assume that the exterior gravitational field of the Sun is well described by the Schwarzschild metric and wish to find observational effects within the Solar System. These effects could test the validity of General Relativity. Doing so requires the study of geodesics, we will derive the geodesic equations using the Lagrangian approached used in Sec. 1.2.5. Our Lagrangian is given by

$$L = -f(r)\dot{t}^2 + \frac{\dot{r}^2}{f(r)} + r^2\dot{\theta}^2 + r^2\sin^2\theta\dot{\phi}^2, \qquad (3.48)$$

with $f(r) = 1 - 2M/r$. The dot stands for the derivative with respect to the geodesic parameter λ. Also, for null geodesics which describe the motion of massless particles like photons we have $L = 0$, while for massive particle we have $L = -1$.

Let us begin with the equation of motion for θ, we have

$$\frac{\partial L}{\partial \theta} = \frac{d}{d\lambda}\frac{\partial L}{\partial \dot{\theta}} \quad \Rightarrow \quad 2\sin\theta\cos\theta\dot{\phi}^2 = 2\frac{d}{d\lambda}(r^2\dot{\theta}). \qquad (3.49)$$

This equation can be solved by choosing $\theta = \pi/2$ which corresponds to aligning the coordinates so that the motion is confined to the equatorial plane. This can always be done for the gravitational two-body problem.

Next, we consider the equation of motion for t. We note that the Lagrangian (3.48) is independent of time, so we expect a constant of motion which we will interpret as energy E. We arrive at

$$\frac{d}{d\lambda}\frac{\partial L}{\partial \dot{t}} = 0 \quad \Rightarrow \quad -f(r)(2\dot{t}) = -2E,\tag{3.50}$$

which means that we can write

$$E = \left(1 - \frac{2M}{r}\right)\dot{t}.\tag{3.51}$$

Similarly, the Lagrangian (3.48) is also independent of the angular variable ϕ so we expect a constant of motion related to angular momentum ℓ for which we have

$$\frac{d}{d\lambda}\frac{\partial L}{\partial \dot{\phi}} = 0 \quad \Rightarrow \quad r^2\dot{\phi} = \ell.\tag{3.52}$$

Substituting the constants of motion E and ℓ back into the Lagrangian yields

$$L = -\frac{E^2}{f(r)} + \frac{\dot{r}^2}{f(r)} + \frac{\ell^2}{r^2}.\tag{3.53}$$

This equation has similarities with a classical mechanical system which can be made more explicit by rewriting

$$\frac{1}{2}\dot{r}^2 + \frac{1}{2}\left(1 - \frac{2M}{r}\right)\left(\frac{\ell^2}{r^2} - L\right) = \frac{1}{2}E^2.\tag{3.54}$$

We can interpret $\dot{r}^2/2$ as the kinetic energy of a test particle with energy $E^2/2$. The remaining term is the effective potential determining the motion of the particle. This effective potential is given by

$$V_{\text{eff}} = \frac{\ell^2}{2r^2} + L\frac{M}{r} - \frac{M\ell^2}{r^3} - \frac{L}{2}.\tag{3.55}$$

Some terms in this effective potential are familiar: $\ell^2/(2r^2)$ corresponds to the centrifugal barrier term, while LM/r is the standard Newtonian term. For massive particles when $L = -1$ this becomes $-M/r$ as expected. The term $-M\ell^2/r^3$ is a new general relativistic term which dominates over the barrier term for small radii.

The earlier-mentioned equations are the starting point for studying geodesics in the Schwarzschild spacetime. We know that the Schwarzschild metric gives a good approximation of the external gravitational field of the Sun and hence we should be able to make some predictions of general relativistic effects which might be observable in the Solar System. This is the subject of Sec. 3.5 and led to the ultimate success of General Relativity.

Before proceeding, we will examine a neat and somewhat counter intuitive example of one particular type of geodesic motion in the Schwarzschild spacetime.

Example 3.1 (Photon sphere). We are interested in the question of whether or not photons (massless particles) can have circular orbits in the Schwarzschild spacetime. For this to be possible, the effective potential V_{eff} must have a stationary point. Setting $L = 0$ in Eq. (3.55), we have

$$V_{\text{eff}} = \frac{\ell^2}{2r^2} - \frac{M\ell^2}{r^3}, \tag{3.56}$$

so that the stationary point $r_\star$, say, is determined by

$$\frac{dV_{\text{eff}}}{dr}(r = r_\star) = -\frac{\ell^2}{r_\star^3} + \frac{3M\ell^2}{r_\star^4} = 0, \tag{3.57}$$

which has the unique physical solution $r_\star = 3M$. This means that at this radius, photons travels on an exact circular trajectory around the central mass. If we looked along the tangential direction at this point, we would see the back of our head in front of us.

A direct calculation also shows that $d^2V_{\text{eff}}/dr^2(r_\star) < 0$ showing that the point $r_\star = 3M$ is a local maximum of the effective potential. This implies that the this point is dynamically unstable. The $r = 3M$ surface of the Schwarzschild spacetime is often referred to

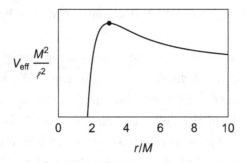

Fig. 3.1   Effective potential V_{eff} given by Eq. (3.56). The dot indicates the position of maximum where $r = r_\star = 3M$.

as the photon sphere. The shape of the potential (3.56) is shown in Fig. 3.1.

Let us now have a closer look at the potential (3.55) for massive particles $L = -1$ so that

$$V_{\text{eff}} = \frac{1}{2}\left(1 - \frac{2M}{r}\right)\left(1 + \frac{\ell^2}{r^2}\right). \qquad (3.58)$$

The shape of this potential is determined by the ratio ℓ^2/M^2, see Fig. 3.2. The possible extremal points of the effective potential are found by solving $dV_{\text{eff}}/dr = 0$ which reduces to a quadratic equation

$$r^2 - \frac{\ell^2}{M}r + 3\ell^2 = 0, \qquad (3.59)$$

and therefore the extremal points are located at

$$r_\pm = \frac{\ell^2}{2M} \pm \frac{\ell^2}{2M}\sqrt{1 - \frac{12M^2}{\ell^2}}. \qquad (3.60)$$

If $12M^2 > \ell^2$, then the potential for massive particles does not have any critical points, for $12M^2 = \ell^2$ there is one point and otherwise there are two points. The smallest possible stationary point for a massive particle is found when setting $12M^2 = \ell^2$, and we obtain $r_{\min} = 6M$ which is twice the radius of the photon sphere. We can

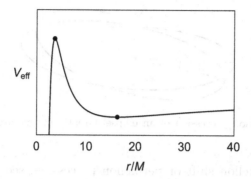

Fig. 3.2 Effective potential V_{eff} for massive particles. The dots indicates the positions of the maximum and minimum.

also solve Eq. (3.59) for the angular momentum ℓ^2 which gives

$$\ell^2 = \frac{rM}{1 - \frac{3M}{r}}, \tag{3.61}$$

and we can use this as an approximation for the angular momentum of an orbit. For an exactly circular stable orbit, this is indeed the correct expression.

3.5. Testing General Relativity — The Classical Tests

In the following we will discuss the three classical tests of General Relativity. These are the perihelion precession of Mercury, the light deflection by the Sun and finally the gravitational redshift of light. We will also discuss a fourth effect, namely gravitational time or radar echo delay which was proposed in the 1960s.

3.5.1. *Perihelion precession of Mercury*

In Newtonian gravity, planets' trajectories around a central object like the Sun are described by exact ellipses. This is highly idealised because the presence of other objects will perturb those trajectories which results in a slight failure of these ellipses to close. This effect

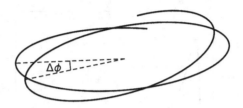

Fig. 3.3   Perihelion precession of an ellipse. Greatly exaggerated for planetary orbits.

is called perihelion shift or perihelion precession, see Fig. 3.3. It is strongest for those objects closest to the central object, and so we are interested in particular in the planet Mercury which is closest to the Sun. The observed perihelion precession for Mercury is about 5,600 arc seconds per century, of which 43 arc seconds cannot be accounted for by Newtonian gravity taking into account all-known perturbative effects.

As we are dealing with the geodesic of a massive particle, we set $L = -1$. To begin with, we need to find an expression for $dr/d\phi$ which can be found by combining Eqs. (3.52) and (3.54) to get

$$\left(\frac{dr}{d\phi}\right)^2 = \left(\frac{\dot{r}}{\dot{\phi}}\right)^2 = \frac{E^2 - \left(1 - \frac{2M}{r}\right)\left(\frac{\ell^2}{r^2} + 1\right)}{\ell^2/r^4}. \qquad (3.62)$$

For problems of this type it is always convenient to introduce a new variable $u = 1/r$ so that Eq. (3.62) becomes

$$\left(\frac{du}{d\phi}\right)^2 = \frac{E^2 - 1}{\ell^2} + \frac{2M}{\ell^2}u - u^2 + 2Mu^3. \qquad (3.63)$$

This differential equation can be solved analytically using elliptic functions. However, since we are only interested in one particular effect described by this equation, there is no need to delve into special functions. The variable u can be considered small (inverse radius) for astrophysical objects. Recall that $-M\ell^2/r^3 = -M\ell^2u^3$ is the new general relativistic term which is cubic in the small quantity u. So, let us study the Newtonian problem first. This means leaving out the cubic term in Eq. (3.63) and considering

$$\left(\frac{du}{d\phi}\right)^2 = \frac{E^2 - 1}{\ell^2} + \frac{2M}{\ell^2}u - u^2. \tag{3.64}$$

We note that we can complete the square on the right-hand side by introducing a new variable $v = u - M/\ell^2$ so that our differential equation simplifies to

$$\left(\frac{dv}{d\phi}\right)^2 = \underbrace{\frac{E^2 - 1}{\ell^2} + \frac{M^2}{\ell^4}}_{=:C^2} - v^2, \tag{3.65}$$

which is of the well-known form $v' = 1/\sqrt{C^2 - v^2}$ and can be solved using separation of variables. A direct calculation gives

$$\frac{1}{r} = v + \frac{M}{\ell^2} = \frac{M}{\ell^2} + C\cos(\phi + \phi_0), \tag{3.66}$$

where ϕ_0 is the constant of integration. This is the equation of an ellipse, which is not obvious at first sight, see Exercise 3.16. However, we do note that this function is 2π periodic.

In order to find the next to leading order correction to this ellipse, we substitute $v = u - M/\ell^2$ into Eq. (3.63) and keep terms up to quadratic order in v instead of u. This results in the new equation

$$\left(\frac{dv}{d\phi}\right)^2 = \frac{E^2 - 1}{\ell^2} + \frac{M^2}{\ell^4} + \frac{2M^4}{\ell^6} + \frac{6M^3}{\ell^4}v - \left(1 - \frac{6M^2}{\ell^2}\right)v^2. \tag{3.67}$$

As before, we can now complete the square on the right-hand side by the substitution

$$w = v - \frac{3M^3}{\ell^4}\left(1 - \frac{6M^2}{\ell^2}\right)^{-1}, \tag{3.68}$$

which transforms Eq. (3.67) in the form

$$\left(\frac{dw}{d\phi}\right)^2 = c_1^2 - c_2^2 w^2, \tag{3.69}$$

and can be integrated using separation of variables. The constant c_2 is simply given by

$$c_2^2 = \left(1 - \frac{6M^2}{\ell^2}\right), \tag{3.70}$$

while the constant c_1 is a more involved expression given by

$$c_1^2 = \frac{E^2 - 1}{\ell^2} + \frac{M^2}{\ell^4} + \frac{2\frac{M^4}{\ell^6} - 3\frac{M^6}{\ell^8}}{\left(1 - \frac{6M}{\ell^2}\right)}. \tag{3.71}$$

Therefore, the solution is given by

$$w = \frac{c_1}{c_2}\cos(c_2\phi + \phi_0), \tag{3.72}$$

where ϕ_0 is a constant of integration. Expressed in our original variable r, the result reads

$$\frac{1}{r} = \frac{M}{\ell^2} + \frac{3M^3}{\ell^4}\left(1 - \frac{6M}{\ell^2}\right)^{-1} + \frac{c_1}{c_2}\cos(c_2\phi + \phi_0). \tag{3.73}$$

In contrast to Eq. (3.66), this solution is not 2π periodic because of the additional factor of c_2 in the argument of cosine. Hence, in General Relativity, the two body problem does not give rise to closed ellipses. There is a slight failure of the object to return to its starting point after one turn.

We can express this by computing the shift of the angle ϕ between two instances where the object is at the same radius r which is

$$\Delta\phi = \frac{2\pi}{c_2} - 2\pi = \frac{2\pi}{\sqrt{1 - \frac{6M^2}{\ell^2}}} - 2\pi. \tag{3.74}$$

In case $c_2 = 1$, we find $\Delta\phi = 0$ and we are back to Newtonian orbits. By assuming that the quantity $M^2/\ell^2 \ll 1$, we can approximate $\Delta\phi$ using a series expansion as follows:

$$\Delta\phi \approx 2\pi\left(1 + \frac{3M^2}{\ell^2}\right) - 2\pi = \frac{6\pi M^2}{\ell^2}. \tag{3.75}$$

Last, we use expression (3.61) for the angular momentum which in first order in M gives $\ell^2 \approx rM$, so that we arrive at

$$\Delta\phi = \frac{6\pi GM}{c^2 r},\tag{3.76}$$

where we re-inserted the physical constants G and c. Next, for the distance of the planet Mercury from the Sun we take the semi-major axis $r = 5.79 \times 10^{10}$ m, while the solar mass is $M_\odot = 1.99 \times 10^{30}$ kg. Note that this result is independent of the mass of Mercury, only its distance from the Sun matters. One computes

$$\Delta\phi = 4.81 \times 10^{-7},\tag{3.77}$$

which is in units of radians per orbit. The orbital period for Mercury is about 0.24 years. Now we convert radians into arc seconds which means we arrive at

$$\Delta\phi = 0.41''/\text{century}.\tag{3.78}$$

One can improve this calculation further, which in particular will take into account the eccentricity of the orbit of Mercury. This results in dividing our result (3.77) by a factor of $(1 - e^2)$ where e is the eccentricity. For Mercury, $e = 0.2$ so that $(1 - e^2) = 0.96$ which yields the improved result

$$\Delta\phi = 0.43''/\text{century},\tag{3.79}$$

which is precisely the amount which cannot be accounted for using Newtonian gravity alone. Amazingly, General Relativity predicts this value for the planet Mercury. Hence, this is a very strong confirmation for Einstein's theory.

3.5.2. *Light deflection by the Sun*

When light or radio signals pass by a massive gravitational object, they will experience a pull towards this object and thus not travel along straight lines, the angle characterising the deviation from a straight line is the deflection angle, see Fig. 3.4. This effect can be

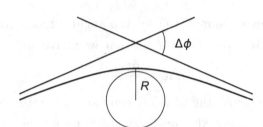

Fig. 3.4 Light deflection by a massive object. Greatly exaggerated for light rays passing by near the Sun.

observed on Earth for signals that pass by near the Sun. For light signals of distant galaxies this can only be done during a Solar eclipse, however, for radio signals this is possible continuously. One compares the observed trajectories with those seen half a year later or earlier when the Sun is no longer between the observer and the source.

We define the deflection angle to be

$$\Delta\phi = \phi_+ - \phi_- - \pi, \tag{3.80}$$

where the subscripts $\pm$ indicate the asymptotic angles, as r becomes very large. The situation is symmetric with respect to the object and hence we can use

$$\Delta\phi = 2\phi_+ - \pi. \tag{3.81}$$

The starting point to compute the deflection angle is again the equation for $d\phi/dr$ which is found by combining Eqs. (3.52) and (3.54) with $L = 0$. This yields

$$\left(\frac{dr}{d\phi}\right)^2 = \frac{E^2 - \left(1 - \frac{2M}{r}\right)\frac{\ell^2}{r^2}}{\ell^2/r^4}. \tag{3.82}$$

The minimum distance r_0 of the signal from the gravitational objects can be defined by solving $dr/d\phi = 0$ for which we find

$$E^2 = \left(1 - \frac{2M}{r_0}\right)\frac{\ell^2}{r_0^2}. \tag{3.83}$$

This follows from setting the numerator in (3.82) to zero. The quantity r_0 is often called the impact parameter in the literature.

Substituting the energy E for the impact parameter, we arrive at the following differential equation for the angle

$$\frac{d\phi}{dr} = \frac{1}{r^2} \left[\left(1 - \frac{2M}{r_0}\right) \frac{1}{r_0^2} - \left(1 - \frac{2M}{r}\right) \frac{1}{r^2} \right]^{-\frac{1}{2}}. \qquad (3.84)$$

Note that this equation is independent of angular momentum ℓ. At this point one could follow an approach similar to that used when computing the perihelion precession. However, one could also use a different technique based on a Taylor series expansion.

One can integrate directly, so that ϕ_+ is given by

$$\phi_+ = \lim_{\bar{r} \to \infty} \int_{r_0}^{\bar{r}} \frac{1}{r^2} \left[\left(1 - \frac{2M}{r_0}\right) \frac{1}{r_0^2} - \left(1 - \frac{2M}{r}\right) \frac{1}{r^2} \right]^{-\frac{1}{2}} dr. \qquad (3.85)$$

First, we introduce a new independent variable $u = 1/r$ for which the integral transform to

$$\phi_+ = \int_0^{1/r_0} \left[\left(1 - \frac{2M}{r_0}\right) \frac{1}{r_0^2} - u^2 - 2Mu^3 \right]^{-\frac{1}{2}} du. \qquad (3.86)$$

Looking at this integral more closely, we see that this is difficult to solve for general r_0 and M. However, as before, we do not need to know the value of this integral exactly, we are mainly interested in the first-order corrections to Newtonian gravity and hence will treat M as a small parameter and make a series expansion of the integrand.

Setting $M = 0$ in Eq. (3.86) gives

$$\phi_+ = \int_0^{1/r_0} \frac{1}{\sqrt{1/r_0^2 - u^2}} du = \arcsin(r_0 u) \Big|_0^{\frac{1}{r_0}} = \frac{\pi}{2}. \qquad (3.87)$$

In the absence of any masses we find $\Delta\phi = 2(\pi/2) - \pi = 0$ so that there is no angle between the incoming ray and the outgoing ray which means that light signals travel along straight lines. This is indeed the expected result for flat space.

For the first-order correction, we need to compute the derivative of the integrand of (3.86) with respect to M and evaluate this derivative for $M = 0$, this leads to

$$\frac{\partial \phi_+}{\partial M}(M = 0) = \int_0^{1/r_0} \frac{1/r_0^3 + u^3}{(1/r_0^2 - u^2)^{3/2}} du. \tag{3.88}$$

One can attack this integral with the substitution $u = \sin(\alpha)/r_0$ which reduces it to a trigonometric integral and can be solved using standard techniques. The final result is given by

$$\frac{\partial \phi_+}{\partial M}(M = 0) = \frac{2}{r_0}, \tag{3.89}$$

and we can write the first two terms of the quantity ϕ_+ as

$$\phi_+ = \frac{\pi}{2} + \frac{2M}{r_0}, \tag{3.90}$$

where the factor of M comes from the Taylor expansion in the mass term. Therefore, the deflection angle of the Schwarzschild spacetime is given by

$$\Delta \phi = 2\phi_+ - \pi = \frac{4M}{r_0}. \tag{3.91}$$

For light rays or radio signals passing nearby the Sun we approximate the distance of closest approach by $r_0 = R_\odot$. We already computed the necessary numbers in Eq. (3.47), which with an additional factor of two yields

$$\Delta \phi = \frac{4GM_\odot}{c^2 R_\odot} = 8.5 \times 10^{-6} = 1.75''. \tag{3.92}$$

This deflection angle was first observed during a solar eclipse in 1919 led by Eddington. It was the first experimental observation of an important prediction by General Relativity.

It appears that we have lost the Newtonian case somewhere along the calculation. When computing the perihelion precession, we first calculated the Newtonian orbit and next the general relativist correction to it. Here, our Taylor series approach gave us the flat space result first and the general relativistic result second. So what is the deflection angle predicted by Newtonian gravity? First, one should ask whether a massless particle like the photon should feel gravitational attraction in the first place. In classical Newtonian physics where the photon is a massless particle, the answer would be No. From a modern physics point of view, on the other hand, one would simply argue that the photon has energy and momentum and hence should feel the gravitational attraction of massive bodies. Once the finiteness of the speed of light was established, one could study the photon path in a Newtonian gravitational field by treating the photon like a real particle moving at the same speed. The resulting bending angle has been known since the late 18th century and is given by

$$\Delta\phi_{\text{Newton}} = \frac{2M}{r_0}, \tag{3.93}$$

which is precisely half the deflection angle predicted by General Relativity.

3.5.3. *Gravitational redshift of light*

Redshift z is a dimensionless quantity used in various fields of physics and is defined by

$$z = \frac{\lambda_{\text{obs}} - \lambda_{\text{e}}}{\lambda_{\text{e}}}, \tag{3.94}$$

where λ_{obs} is the wavelength of the observed signal and λ_{e} is the wavelength of the emitted signal. This redshift will play an important role in Cosmology, see Secs. 4.3.2 and 4.3.4.

Let us consider a signal emitted with wavelength λ_{e}, then this signal will have local energy hf_{e} where h is Planck's constant and f_{e} is the frequency of the photon (recall $\lambda f = c$). We discussed in Sec. 2.1.2 the proper time of a local observer. The change of the proper time of the emitter determines the frequency of the emitted wave while the proper time of the observer determines the frequency of the observed wave.

In General Relativity, the strength of the gravitational field will affect clock cycles. Let us consider an emitter and an observer, both of which are at fixed spatial coordinates, for instance we can think of the Sun emitting some radiation which is observed on Earth. The local movement of the Earth relative to the Sun during the wave's propagation will introduce an additional Doppler effect, however, for now we are interested in the purely gravitational effect. Let us consider the Schwarzschild metric (3.19), and assume the emitter is located at r_e, θ_e, ϕ_e, then the proper time is given by

$$dr_e = \sqrt{-g_{ij}dX^idX^j} = \sqrt{1 - \frac{2M}{r_e}}dt_e. \tag{3.95}$$

Likewise, the proper time of the observer is

$$d\tau_{obs} = \sqrt{1 - \frac{2M}{r_{obs}}}dt_{obs}. \tag{3.96}$$

Before proceeding, let us briefly compare these two equations with the time dilation relation (2.20), we write

$$dt = \frac{1}{\sqrt{1 - \frac{2M}{r}}}d\tau, \tag{3.97}$$

for some fixed radius r and note that the term from the Schwarzschild metric plays the role of the Lorentz factor. Hence, clocks near massive objects will also be running slower than clocks far away where the gravitational field is weak.

We can interpret $d\tau_e$ and $d\tau_{obs}$ as the times between two maxima of the emitted and observed waves, respectively. Combining

Eq. (3.94) with (3.95) and (3.96) gives

$$z = \frac{d\tau_{\text{obs}}}{d\tau_e} - 1 = \sqrt{\frac{1 - \frac{2M}{r_{\text{obs}}}}{1 - \frac{2M}{r_e}}} - 1, \qquad (3.98)$$

where we note that the coordinate time differences dt_e and dt_{obs} are equal for spatially fixed emitters and observers.

Let us now consider again the Sun, and estimate the redshift z for a signal emitted on the surface of the Sun and received on Earth. Since the distance between the Sun and the Earth is much larger than the radius of the Sun, we can neglect the term $2M_\odot/r_{\text{obs}} \ll 1$ in Eq. (3.98), therefore

$$z \approx \frac{1}{\sqrt{1 - \frac{2M_\odot}{R_\odot}}} - 1 \approx \frac{M_\odot}{R_\odot}, \qquad (3.99)$$

where in the final step we assumed $M_\odot/R_\odot$ to be small. Taking the physical mass and radius of the Sun and reinserting the gravitational constant and the speed of light yields

$$z = \frac{GM_\odot}{c^2 R_\odot} \approx 2 \times 10^{-6}. \qquad (3.100)$$

This effect is fairly small which makes it difficult to measure. Due to the high surface temperature of the Sun, the thermal velocities of various atoms are of the order of 10^3 m/s, this will introduce Doppler effects which are of similar order of magnitude to the redshift z. Nonetheless, observational data is in excellent agreement with the predictions of General Relativity.

Let us make a small remark about the gravitational redshift. The discussion is independent of the Einstein field equations, we would have arrived at the same result using only Eqs. (2.59) and (2.66). In this sense, the gravitational redshift is not directly testing General Relativity but it is verifying the principle of equivalence.

3.5.4.   *Radar echo or gravitational time delay*

Strictly speaking the gravitational time delay is not one of the three classical tests, however, the idea behind this test is very similar to the earlier-mentioned ones where a signal passes nearby the Sun. An intense electromagnetic wave is directed to another planet when this planet is almost opposite to the Earth on the far side of the Sun. A radio telescope on Earth is detecting the reflection or echo of this signal. As the signal and its reflection pass by near the Sun, the photons will not travel along straight lines. Hence, their travel times will differ from those expected from a straight line path. Now we will estimate this time delay for Earth and Venus, but it equally applies to other planets.

The starting point of this calculation is the geodesic equation with $L = 0$ for photons. We are interested in the coordinate time t, so combining Eq. (3.51) with Eq. (3.53) gives an expression for dr/dt given by

$$\left(\frac{dr}{dt}\right)^2 = \left(1 - \frac{2M}{r}\right)^2 - \frac{\ell^2}{E^2 r^2}\left(1 - \frac{2M}{r}\right)^3. \qquad (3.101)$$

As in Sec. 3.5.2 we introduce the impact parameter r_0 as the distance of closest approach of the photon to the Sun. This minimum distance is defined by $dr/dt(r_0) = 0$, so that we find

$$\frac{E^2}{\ell^2} = \frac{1}{r_0^2}\left(1 - \frac{2M}{r_0}\right), \qquad (3.102)$$

which can be substituted back into Eq. (3.101) to eliminate E^2/ℓ^2. This yields

$$\frac{dr}{dt} = \left(1 - \frac{2M}{r}\right)\left[1 - \frac{r_0^2}{r^2}\frac{1 - 2M/r}{1 - 2M/r_0}\right]^{\frac{1}{2}}. \qquad (3.103)$$

Now we can separate the variables and integrate to find the travel time between some radius r_1 and the distance of closest approach r_0, which is

$$t(r_0, r_1) = \int_{r_0}^{r_1} \left(1 - \frac{2M}{r}\right)^{-1} \left[1 - \frac{r_0^2}{r^2} \frac{1 - 2M/r}{1 - 2M/r_0}\right]^{-\frac{1}{2}} dr. \quad (3.104)$$

Many integrals that appear when studying geodesics in the Schwarzschild spacetime need to be approximated since one cannot find explicit solutions in terms of elementary functions in closed form. As in the previous sections, we assume M to be small. Then the integrand in first order in M is given by

$$\frac{r}{\sqrt{r^2 - r_0^2}} \left[1 + \frac{3M}{r} - \frac{M}{r + r_0}\right], \quad (3.105)$$

and the three resulting terms are standard integrals which lead to the result

$$t(r_0, r_1) = \sqrt{r_1^2 - r_0^2} + M\sqrt{\frac{r_1 - r_0}{r_1 + r_0}} + 2M \log\left(\frac{r + \sqrt{r^2 - r_0^2}}{r_0}\right). \quad (3.106)$$

The first term corresponds to the travel time of the signal had it followed a straight line path. The terms proportional to the mass contain the additional travel time due to the curved trajectory of the photon, this excess time between two points is therefore given by $t(r_0, r_1) - \sqrt{r_1^2 - r_0^2}$. Let us now consider a signal sent from Earth ♁ to Venus ♀, then reflected back to Earth where it is observed. The excess time compared to the straight line is given by

$$\Delta t = 2\left[t(r_♁, r_0) - \sqrt{r_♁^2 - r_0^2} + t(r_♀, r_0) - \sqrt{r_♀^2 - r_0^2}\right], \quad (3.107)$$

where the factor of 2 comes from the fact that first the signal and then its reflection have to be considered. In simple words, the photon has to travel to Venus and then back again.

For signals which pass nearby the Sun, we can assume $r_0 \ll r_\oplus$ and $r_0 \ll r_♀$, which allows us to further simplify the time delay. In lowest order in r_0 we find

$$c\Delta t \approx \frac{4GM}{c^2} \left[1 + \log \left(\frac{r_\oplus r_♀}{r_0^2} \right) \right], \qquad (3.108)$$

where we reinserted the physical constants G and c. In principle, we should take into account that the proper time $\Delta\tau$ for the signal differs from the coordinate time Δt, however, since $2GM_\oplus/c^2/r_\oplus \ll 1$ this effect can be ignored in the present calculation. We obtain the final result

$$\Delta t = \frac{4GM_\odot}{c^3} \left[1 + \log \left(\frac{r_\oplus r_♀}{R_\odot^2} \right) \right] \approx 220 \times 10^{-6}\text{s}. \qquad (3.109)$$

While there are some technicalities that somewhat limit the accuracy of measuring this effect, it is fair to say that measurements are again in excellent agreement with the prediction by General Relativity.

3.6. The Schwarzschild Radius

When we solved the Einstein field equations to find the Schwarzschild solution, we already noted that this metric is singular when $r = 2M$ or $r = 0$. Our first aim is to understand whether the $r = 2M$ surface corresponds to a real physical singularity or to choosing 'bad' coordinates to cover the manifold. Clearly, when we solved the field equations, we were primarily interested in finding a solution in some coordinate system which took into account our requirements of staticity and spherical symmetry, rather than worrying about the entire manifold.

3.6.1. *Radial null geodesics*

Let us briefly recall Examples 1.10 and 1.13. In both cases the geodesics of these spaces proved invaluable to understand their geometric properties. Therefore, it seems natural to consider radial geodesics of the Schwarzschild spacetime. Radial geodesics have no angular momentum, so we set $\ell = 0$, and moreover we will be primarily interested in radial null geodesics which means $L = 0$. In more physical terms we are studying the propagation of radial photons. The geodesic equations are (3.51) and (3.54) which become

$$E = \left(1 - \frac{2M}{r}\right)\dot{t}, \tag{3.110}$$

$$\dot{r}^2 = E^2. \tag{3.111}$$

The second equation becomes $\dot{r} = \pm E$ and can be integrated with respect to the affine parameter λ. This gives

$$r = r_0 \pm E\lambda, \tag{3.112}$$

where r_0 is a constant of integration which corresponds to the distance of the photon from the centre when $\lambda = 0$. For the positive sign the radius increases with λ and hence we will speak of outgoing null geodesics while for the negative sign the radius decreases with λ and so we will speak of incoming null geodesics. The most notable implication of Eq. (3.112) is that the $r = 2M$ surface does not appear to introduce any conceptual problems when dealing with radial null geodesics. Therefore, any such geodesic will reach the Schwarzschild radius at finite affine parameter. However, this changes quite dramatically when working with coordinate time. We have

$$\frac{dt}{dr} = \frac{\dot{t}}{\dot{r}} = \pm \left(1 - \frac{2M}{r}\right)^{-1}, \tag{3.113}$$

for outgoing and incoming geodesics, respectively. This can be integrated with respect to r and we arrive at

$$t - t_0 = \pm(r + 2M \log |r - 2M|), \qquad (3.114)$$

where t_0 is a constant of integration. Hence, as $r \to 2M$, the coordinate time t approaches positive or negative infinity for outgoing or incoming null geodesics, respectively, in contrast to the finite affine parameter.

3.6.2. *Eddington–Finkelstein coordinates*

We can solve Eq. (3.114) for the constant t_0 for both signs and arrive at quantities which are constant for either outgoing or incoming null geodesics. We define

$$u = t - r - 2M \log \left| \frac{r}{2M} - 1 \right|, \qquad (3.115)$$

$$v = t + r + 2M \log \left| \frac{r}{2M} - 1 \right|, \qquad (3.116)$$

where u is constant for the outgoing null geodesics and v is constant for the incoming ones. We can visualise the incoming and outgoing geodesics in the original Schwarzschild coordinates (t, r) in Fig. 3.5.

It is now tempting to introduce new coordinates for the Schwarzschild spacetime based on u and v. These are called the Eddington–Finkelstein coordinates, u is the outgoing Eddington–Finkelstein coordinate and v is the incoming Eddington–Finkelstein coordinate. Let us eliminate the time coordinate t using the new coordinate u, we have

$$dt = du + \frac{r}{r - 2M} dr, \qquad (3.117)$$

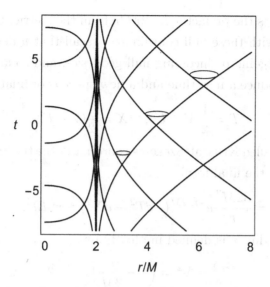

Fig. 3.5 Incoming and outgoing geodesics in Schwarzschild coordinates (t, r), the lines correspond to $u = $ const. and $v = $ const., the Schwarzschild radius $r = 2M$ and the origin $r = 0$ are indicated by thick lines. Also included are some light cones.

so that the Schwarzschild metric (3.19) in outgoing Eddington–Finkelstein coordinates becomes

$$ds^2 = -\left(1 - \frac{2M}{r}\right) du^2 - 2dudr + r^2 d\Omega^2. \qquad (3.118)$$

In these coordinates the Schwarzschild metric is no longer singular at $r = 2M$, however, the metric remains singular at $r = 0$. A similar form of the metric is found when working with ingoing Eddington–Finkelstein coordinates.

3.6.3. *Kruskal–Szekeres coordinates*

In order to find a coordinate system which covers the entire manifold, we first introduce another set of coordinates given by

$$U = -e^{-u/4M}, \quad V = -e^{v/4M}, \qquad (3.119)$$

which removes the prefactor $1 - 2M/r$ from the metric. Last, instead of working with these null coordinates (recall that u and v are constant for outgoing or incoming null geodesics, and so are U and V) we will introduce a new time and a new space coordinate as follows:

$$T = \frac{1}{2}(V + U), \quad X = \frac{1}{2}(V - U). \tag{3.120}$$

In these so-called Kruskal–Szekeres coordinates the the Schwarzschild metric takes the final form

$$ds^2 = \frac{32M^3}{r}e^{-r/2M}(-dT^2 + dX^2) + r^2 d\Omega^2, \tag{3.121}$$

where the radius r is defined implicitly by

$$T^2 - X^2 = \left(1 - \frac{r}{2M}\right)e^{-r/2M}. \tag{3.122}$$

Metric (3.121) is now well defined and regular for all $r > 0$. Therefore, we can create a spacetime diagram for the entire manifold, see Fig. 3.6.

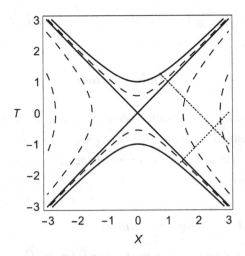

Fig. 3.6 Incoming and outgoing geodesics in Kruskal–Szekeres coordinates (T, X), the dotted diagonal lines correspond to the incoming and outgoing null geodesics. The Schwarzschild radius $r = 2M$ and the origin $r = 0$ are indicated by thick lines. Some $r = $ const. lines are indicated by dashed lines.

3.6.4. *Black holes*

After discussing geodesics in the Schwarzschild spacetime, we are now in a position to be slightly more precise about black holes. The Schwarzschild radius is often referred to as the event horizon of black holes as it separates two regions which are of particular interest. From Fig. 3.6 we conclude that an outgoing null geodesic emitted at any radius r which is between the event horizon and the centre $0 < r < 2M$ cannot leave this region. Graphically this follows from the fact that radial null geodesics travel along diagonal lines and hence cannot intersect with the $r = 2M$ lines which are also diagonals. On the other hand, all incoming radial null geodesics will eventually cross the horizon and approach the singularity at $r = 0$. The precise mathematical definition of a black hole in General Relativity is in fact not as simple as it appears as there is no reason to believe that black holes have to be spherically symmetric. Since most astrophysical objects are rotating, we would expect a black hole to also rotate and hence have angular momentum, see for instance Wald (1984). We mentioned that the centre of the black hole corresponds to a true spacetime singularity which is 'hidden' from the outside by the black hole horizon. It is widely believed that the gravitational collapse of stars or other matter sources will produce black holes instead of 'naked singularities' which would be a singularity without a horizon hiding it. Research in these directions is ongoing and some of the questions involved are very subtle.

3.7. Further Reading

This entire chapter focussed on one special solution of the Einstein field equations. It would therefore be natural to continue with other known exact solutions to the field equations. Of particular interest is the Kerr solution which we can view as the rotating generalisation of the Schwarzschild solution, see for instance Stephani *et al.* (2003, Chap. 20). Recall that the Schwarzschild solution was found only a year after the field equations were discovered. The Kerr solution, on the other hand, was only found in 1963. The axially-symmetric Einstein field equations are considerably more difficult than the spherically symmetric ones. There are many other interesting solutions of the field equations which could form the basis of further studies, on the more speculative end this would include wormhole solutions for instance.

An interesting gravitational effect due to the rotation of a massive object like the Earth is the so-called Lense–Thirring effect, sometimes called frame-dragging effect. This very small effect would change the spin direction of gyroscopes that orbit the Earth. This effect was eventually confirmed experimentally by Gravity Probe B in 2011. The experimental verification of General Relativity is thoroughly discussed in Will (2014).

Black holes are also a fascinating subject of current research. One can show that the total area of all black holes in the Universe cannot decrease which shows strong similarities with the second law of thermodynamics. This analogy goes much further and one can formulate analogues of the other thermodynamic laws in the context of black holes, see Wald (1984). In 1974, Hawking used a quantum field theoretical calculation to show that black holes emit black body radiation, Hawking radiation, of a specific temperature, the so-called Hawking temperature.

The previously mentioned centennial perspective by Ashtekar *et al.* (2015) would also be a suitable continuation for readers interested in an overview of current research topics. Large parts of this book should be accessible to readers at this point, all articles contain references to the original literature.

For readers who wish to improve their problem and exercise-solving skills, probably the best book available is by Lightman *et al.* (1975). It contains almost 500 problems with fully worked-out solutions.

The recommended literature is by no means complete, it simply reflects the author's suggestions for further reading.

3.8. Exercises

The Schwarzschild Solution

Exercise 3.1. Find the Schwarzschild radius of a proton and compare this with its actual radius.

Exercise 3.2. Find the non-vanishing Christoffel symbol components of metric (3.5).

Exercise 3.3. Show that Eq. (3.22) is indeed equivalent to Eq. (3.19).

Exercise 3.4 (takes time). Consider the static and spherically symmetric metric $ds^2 = -e^{2A(r)}dt^2 + e^{2B(r)}(dr^2 + r^2 d\theta^2 + \sin^2\theta d\phi^2)$ in isotropic coordinates. Recall that this r is different from the radial coordinate r used in standard Schwarzschild coordinates. Compute all non-vanishing Christoffel symbol components and then compute the non-vanishing Ricci tensor components.

Exercise 3.5. Use the Ricci tensor components from the previous exercise to find the non-vanishing components of the Einstein tensor.

Exercise 3.6 (takes time). Finally, continuing from the previous two exercises solve the vacuum field equations, thereby deriving from first principles the Schwarzschild solution in isotropic coordinates given by Eq. (3.22).

Exercise 3.7 (hard). Consider the metric

$$ds^2 = -dt^2 + \left(\frac{\mu/3}{r-t}\right)^{\frac{2}{3}} dr^2 + \left(\frac{9\mu}{8}(r-t)^2\right)^{\frac{2}{3}} d\Omega^2.$$

Show that this is the Schwarzschild metric by finding the coordinate transformation which transforms this metric into the form (3.19). Find the relationship between μ and mass parameter M.

Exercise 3.8. Using the Schwarzschild coordinates (3.5), solve the vacuum field equations in the presence of the cosmological term. The result is the so-called Schwarzschild–de Sitter or Kottler solution which is given by

$$ds^2 = -\left(1 - \frac{2M}{r} - \frac{\Lambda}{3}r^2\right) dt^2 + \left(1 - \frac{2M}{r} - \frac{\Lambda}{3}r^2\right)^{-1} dr^2 + r^2 d\Omega^2.$$

The Schwarzschild Interior Solution

Exercise 3.9. The spatial part of the Schwarzschild interior metric (3.33) can be written in the form $ds^2 = dr^2/(1 - kr^2) + r^2 d\Omega^2$. Show that this is the metric of a 3-sphere.

Exercise 3.10. By solving the field equation with components $i = j = 0$ in Eq. (3.6), find the spatial part of the Schwarzschild interior metric with cosmological constant. Assume constant density.

Exercise 3.11. For the Schwarzschild interior solution we introduced the mass definition Eq. (3.29) which corresponds to the Newtonian mass. This definition ignores the fact that we should integrate on a curved manifold and also ignores that pressure contributes to the gravitational field. The proper mass is defined by

$$M_p = \int_0^R 4\pi\rho(\bar{r})\bar{r}^2 \frac{d\bar{r}}{\sqrt{1 - 2m(\bar{r})/\bar{r}}}. \tag{3.123}$$

Find the proper mass for the Schwarzschild interior solution.

Exercise 3.12 (hard). The mass measured at infinity of a static and spherically symmetric star is given by

$$M_\infty = \int_0^R (\rho + 3p)e^{A/2+B/2}4\pi\bar{r}^2 d\bar{r}. \tag{3.124}$$

Show that this mass equals to total mass $M = m(R)$ defined by Eq. (3.31).

Geodesics in Schwarzschild Spacetime

Exercise 3.13 (A classic). Consider a radio commentator falling radially into a Schwarzschild black hole. As he approaches the Schwarzschild radius, his broadcast wavelength strongly redshifts. The radio listener (far away from the black hole) observes the time dependence of this redshift $\lambda_{\text{obs}}/\lambda_e \approx \exp(-t/\mu)$. Find the relationship between μ and the mass of the black hole.

Exercise 3.14. The setting of this question is the Schwarzschild spacetime. Consider a freely falling, massive test particle initially at rest at $r = 10M$. Show that the proper time required for this particle to reach the centre $r = 0$ is given by $5\sqrt{5}M\pi$.

Exercise 3.15 (hard). The de Sitter solution in Schwarzschild coordinates is given by

$$ds^2 = -\left(1 - \frac{\Lambda}{3}r^2\right) dt^2 + \left(1 - \frac{\Lambda}{3}r^2\right)^{-1} dr^2 + r^2 d\Omega^2, \tag{3.125}$$

which is the $M \to 0$ limit of the Kottler or Schwarzschild–de Sitter solution. Show that for $\Lambda \leq 0$ this metric is regular everywhere. Identify the radius r_Λ where the metric is singular when $\Lambda > 0$.

Find the geodesic equations of the de Sitter solution and consider radial geodesics. Show that a freely falling observer starting at the origin with velocity v will cross the surface $r = r_\Lambda$ for finite affine parameter, thereby showing that r_Λ corresponds to a coordinate singularity.

Testing General Relativity — The Classical Tests

Exercise 3.16 (hard). Show that Eq. (3.66) is indeed the equation of an ellipse in this context. Determine the relationship between the constants in this equation and those characterising the ellipse.

Exercise 3.17. Estimate the deflection angle of light rays or radio signals passing nearby the surface of Jupiter, and compare it to that of the Sun.

Exercise 3.18. Consider the integrand of Eq. (3.104) and make a series expansion in the mass parameter M up to linear terms.

The Schwarzschild Radius

Exercise 3.19. Consider the de Sitter metric (3.125). Introduce a new coordinate $u = t - f(r)$ which replaces the time coordinate. Find $f(r)$ such that the de Sitter metric becomes regular across r_Λ. Is this metric regular everywhere?

Exercise 3.20. Derive the Schwarzschild metric in incoming Eddington–Finkelstein coordinates.

Exercise 3.21. Derive the Schwarzschild metric using both incoming and outgoing Eddington–Finkelstein coordinates (u, v) instead of the coordinates (t, r).

Exercise 3.22. Using Eq. (3.120), find the explicit coordinate transformation for the Kruskal–Szekeres coordinates T and X in terms of the original Schwarzschild coordinates t and r. Distinguish between $r > 2M$ and $r < 2M$ (this part is harder).

4

Cosmology

Cosmology is the study of the universe as a whole. Its study has a very long history, humanity at all times was interested in the universe and longed to understand it. Physical cosmology in its modern form began shortly after the formulation of General Relativity when various researchers were interested in understanding the cosmological consequences of the theory.

4.1. Classical and Modern Cosmology

4.1.1. *Cosmological principle*

One can broadly differentiate between classical cosmology and modern cosmology. Classical cosmology focusses on the study of particular solutions of the Einstein field equations which could model the universe. It was in 1929 when Hubble observed the redshift of distant galaxies thereby providing observational evidence for an expanding universe. The next crucial observation was the discovery of the cosmic microwave background radiation by Penzias and Wilson in 1964. At this point in time cosmology was still a relatively small research field, often seen as the more speculative end of General Relativity.

While the 1964 observation was hugely important, it was much later that cosmology was transformed into a substantial and independent research field. In 1998/1999, observations of type Ia supernovae showed that the universe is not just expanding, but that this expansion is in fact accelerating. The most straightforward explanation for such a behaviour is to introduce the cosmological constant Λ into the Einstein field equations. Together with the COBE (1989) and WMAP (2001) missions, which studied in detail the structure of the anisotropies of the cosmic microwave background radiation, the so-called standard model of cosmology was formulated. Modern Cosmology generally refers to the study of this standard model. In the following we will discuss the most important aspects of Classical Cosmology and then introduce the concepts of Modern Cosmology.

We recall that the Einstein field equations are very complicated in general. However, when studying the static and spherically symmetric case in Sec. 3.1, it turned out that the field equations simplified considerably because of this symmetry. Our first task therefore is to identify some suitable assumptions about the universe so that we arrive at manageable field equations.

In order to do this, we also need to make some fundamental assumptions about the laws of physics within our universe. We are unlikely to be able to directly test the validity of physical laws on cosmological scales, we also need to establish what this scale is. As our primary working assumption we assume that the laws of physics are the same everywhere and at all times, and that the universe is connected.

Let's imagine a night sky and we are looking in a certain direction. We note that some areas contain more brighter objects than others. We are taking this further by placing a fictitious cube with side length ℓ into any part of our observable universe and count the number of galaxies in this volume. It turns out that the number of galaxies in any cube of side length of about $100\,\text{Mpc}$ ($1\,\text{pc} \approx 3.26\,\text{ly} \approx 3.09 \times 10^{16}\,\text{m}$) is approximately the same. On such scales the universe looks statistically uniform, or homogeneous. On the other hand, the cosmic microwave background radiation has roughly the same intensity in every direction, it is isotropic. These two observation are generally promoted to what is known as the cosmological principle.

Cosmological principle: *The universe is homogeneous and isotropic at all times when viewed on large enough scales.*

This means the universe is the same for all observers, independent of their location. On cosmological scales, test particles and observers are galaxies modelled as a perfect fluid with energy density ρ and isotropic pressure p. By a cosmological reference frame we mean a set of coordinates in which physical quantities are homogeneous and isotropic. In this reference frame we consider a comoving observer which is at rest in that frame. The proper time t measured by this comoving observer, starting at $t = 0$, is called the cosmological or cosmic time. This cosmological time is very similar to Newtonian time in classical mechanics.

4.1.2. *Geometry of constant time hypersurfaces*

At any particular cosmic time t_1 the universe defines a 3D manifold, a space-like hypersurface. It turns out that homogeneity and isotropy imply that this hypersurface is a space of constant curvature. Any two such spaces of the same dimension, same metric signature and same value of constant curvature are locally isometric. Fortunately, spaces of constant curvature are quite simple. Spaces of constant positive curvature are spheres, space with vanishing curvature are Euclidean spaces and spaces of constant negative curvature are hyperbolic spaces.

It can be shown that the metric of a 3D space of constant curvature can be written as

$$\gamma_{ij} dX^i dX^j = \frac{dx^2 + dy^2 + dz^2}{\left(1 + \frac{k}{4}(x^2 + y^2 + z^2)\right)^2}, \qquad (4.1)$$

where x, y, z are the standard Euclidean spatial coordinates, and k is the curvature constant which can take the values $k = \{+1, 0, -1\}$. The Ricci scalar of this 3-metric is given by

$$R_\gamma = 6k, \qquad (4.2)$$

which is a constant, and hence, the sign of k determines the sign of the scalar curvature. Recall that we encountered various spaces of constant curvature in the previous sections. Exercise 1.11 discussed hyperbolic space, while Exercise 3.9 was about the 3-sphere.

Working with standard Euclidean coordinates in Eq. (4.1) has some advantages when it comes to interpreting results. However, this choice of coordinates if somewhat inconvenient since the denominator depends on all spatial coordinates. To begin with we introduce standard spherical polar coordinates (1.52)–(1.54). In these coordinates the metric becomes

$$\gamma_{ij} dX^i dX^j = \frac{dr^2 + r^2 d\Omega^2}{\left(1 + \frac{k}{4}r^2\right)^2}, \tag{4.3}$$

and r is the Euclidean distance from the origin. This metric simplifies considerably if we introduce the new radial coordinate

$$\rho = r \left(1 + \frac{k}{4}r^2\right)^{-1}. \tag{4.4}$$

In the coordinates $X^i = \{\rho, \theta, \phi\}$ metric Eq. (4.1) takes the form

$$\gamma_{ij} dX^i dX^j = \frac{d\rho^2}{1 - k\rho^2} + \rho^2 d\Omega^2, \tag{4.5}$$

which is easier to handle for explicit computations.

Note. Many authors, including myself, often write metric (4.5) using r instead of ρ for the radial coordinate. For the purpose of this text it is cleaner to use r for the true Euclidean distance and use ρ to emphasise that we are working with a different radial coordinate, as defined in Eq. (4.4).

When $k = 1$, metric (4.5) is that of a 3-sphere (see Exercise 3.9) which has finite volume. Recall that the volume of the manifold is given by Eq. (1.57) which allowed us to show that $V(\mathbb{S}^3) = 2\pi^2$. The respective volumes of Euclidean space and hyperbolic space are infinite.

4.1.3. *Friedmann–Lemaître–Robertson–Walker metric*

The discussions of the previous subsection are now combined into a metric ansatz which allows us to study homogeneous and isotropic cosmological models in the context of General Relativity. This so-called Friedmann–Lemaître–Robertson–Walker metric is given by

$$ds^2 = -dt^2 + a(t)^2 \left[\frac{d\rho^2}{1 - k\rho^2} + \rho^2 d\Omega^2 \right], \qquad (4.6)$$

where the unknown function $a(t)$ is called the scale factor or sometimes expansion parameter. The name scale factor is very natural as this function 'scales the size' of the spatial part of that metric. An obvious question is whether one can generalise this metric by working with an additional function $N(t)$ so that the temporal part of the metric becomes $-N(t)^2 dt^2$. In fact, this is equivalent to Eq. (4.6) since we can always introduce a new time coordinate t' via $dt' = N(t)dt$ which will absorb the new function. We noted that Eq. (4.5) uses a convenient choice of coordinates, however, one can introduce yet another coordinate system which is particularly elegant and often used.

For $k = 1$, we introduce a third angle by $\rho = \sin \chi$, when $k = 0$ we simply relabel $\rho = \chi$. When $k = -1$ we introduce a hyperbolic angle $\rho = \sinh \chi$. A direct calculation shows that metric (4.6) can now be written as follows:

$$ds^2 = -dt^2 + a(t)^2 \left[d\chi^2 + \Sigma(\chi)^2 d\Omega^2 \right]. \qquad (4.7)$$

The function $\Sigma(\chi)$ is given by

$$\Sigma(\chi) = \begin{cases} \sin \chi & \text{if } k = +1, \\ \chi & \text{if } k = 0, \\ \sinh \chi & \text{if } k = -1. \end{cases} \qquad (4.8)$$

One can also combine these three cases into one function by writing

$$\Sigma(\chi) = \frac{\sin\left(\sqrt{k}\chi\right)}{\sqrt{k}}, \tag{4.9}$$

provided one carefully deals with the limit $k \to 0$, and also recalls the relationship between trigonometric and hyperbolic functions, namely $\sin(ix) = i\sinh(x)$. The quantity $\Sigma(\chi)$ can be thought of as the metric distance in this space and hence we will also write $d_{\text{metric}} = \Sigma(\chi)$ when discussing cosmological distances.

There are two nice things about this function. First, it somewhat justifies why one speaks of spherical and hyperbolic geometries (be careful here, in general one cannot deduce geometrical facts directly from the metric as it could be a simple space written in very strange coordinates). Second, let us assume that $\chi \ll 1$ and write the series expansion of $\Sigma(\chi)$, we find

$$\Sigma(\chi) = \chi + O(\chi^3). \tag{4.10}$$

We now see one of the key features of differential geometry, locally all spaces appear to be flat, and that curvature is a higher-order effect.

Another important remark is in order at this point. Even if $k = 0$ so that the spatial geometry is Euclidean, the full 4D manifold is not flat. We will see this more explicitly when discussing the field equations where we will state the Ricci scalar which is non-zero for $k = 0$.

4.1.4. *Particle horizons*

When discussing cosmology in the context of General Relativity we need to ask the rather natural question: How much of the universe can we in principle observe? Since signals can only travel at the speed of light, the age of the universe multiplied by the speed of light gives us a certain length scale which restricts what we can potentially observe at any given point in cosmological time. In models where the universe is expanding in time, this issue is even more pressing.

In order to simplify the following discussion, we will briefly restrict ourselves to $k = 0$. In this case we write our cosmological metric as

$$ds^2 = -dt^2 + a(t)^2 \left(dx^2 + dy^2 + dz^2\right), \tag{4.11}$$

and work with spatial Euclidean coordinates. Make the coordinate transformation

$$\tau = \int \frac{dt}{a(t)}, \quad d\tau = \frac{dt}{a(t)}, \tag{4.12}$$

so that Eq. (4.11) transforms into

$$ds^2 = a(\tau)^2 \left(-d\tau^2 + dx^2 + dy^2 + dz^2\right). \tag{4.13}$$

The metric in this form is a multiple of Minkowski space and τ is called conformal time. Hence, we would expect all coordinates to range from $-\infty$ to $+\infty$. However, for τ being able to take all such values, the integral in Eq. (4.12) cannot be arbitrary which in turn implies that $a(t)$ is somewhat restricted. Let us make this more formal.

We would like to establish which observers could have sent a signal which reaches another observers at or before an event p. The boundary between those world lines that can reach the second observer and those that cannot is called the particle horizon at p. This is a local definition as different observers will have different particle horizons.

Let us assume the universe began at $t = 0$ and let us consider an observer at some time t_{obs}, which means we are considering a universe of finite age. If τ diverges as $t \to 0$, the observer will be able to receive signals from each and every point of the observer's past. If $a(t) = a_0 t^\alpha$ with a positive constant a_0, then the integral in Eq. (4.12) diverges as $t \to 0$ if $\alpha \geq 1$ and hence there will be no particle horizon. If, on the other hand, the integral converges, $\alpha < 1$, then there exits a particle horizon because τ will be bounded from below and so part of the spacetime described by (4.13) is 'missing'.

4.1.5. *Field equations*

The cosmological Einstein field equations are

$$G_{ij} + \Lambda g_{ij} = 8\pi T_{ij}, \tag{4.14}$$

where Λ is the cosmological constant. Its value is about $\Lambda \approx 10^{-52} \mathrm{m}^{-2}$ which means that it is only of relevance on very large scales. We can use the radius associated with the cosmological constant as a rough guide to the size of the universe, we have

$$r_\Lambda = \frac{1}{\sqrt{\Lambda}} \approx 10^{26} \mathrm{m}, \tag{4.15}$$

which is in good agreement with various observations.

The Ricci scalar for our metric (4.6) is given by

$$R = 6\frac{\ddot{a}}{a} + 6\frac{\dot{a}^2}{a^2} + 6\frac{k}{a^2}, \tag{4.16}$$

where the dots denote differentiation with respect to cosmological time t. The components of the Einstein tensor are given by

$$G^0_0 = -3\frac{\dot{a}^2}{a^2} - 3\frac{k}{a^2}, \tag{4.17}$$

$$G^1_1 = G^2_2 = G^3_3 = -2\frac{\ddot{a}}{a} - \frac{\dot{a}^2}{a^2} - \frac{k}{a^2}. \tag{4.18}$$

Likewise, the non-vanishing components of the energy–momentum tensor of a perfect fluid are

$$T^0_0 = -\rho, \quad T^1_1 = T^2_2 = T^3_3 = p, \tag{4.19}$$

where ρ stands for the energy–density and p is the pressure.

The cosmological Einstein field equations are often written in two different but equivalent forms. The first form is the one directly obtained by equating the Einstein tensor with the matter tensor and adding the cosmological term, which results in

$$-3\frac{\dot{a}^2}{a^2} - 3\frac{k}{a^2} + \Lambda = -8\pi\rho, \tag{4.20}$$

$$-2\frac{\ddot{a}}{a} - \frac{\dot{a}^2}{a^2} - \frac{k}{a^2} + \Lambda = 8\pi p. \tag{4.21}$$

An alternative form of these equations which is often used in cosmology is found by eliminating the term $\dot{a}/a$ from the second equation (4.21) which yields

$$\frac{\dot{a}^2}{a^2} = \frac{8\pi\rho}{3} + \frac{\Lambda}{3} - \frac{k}{a^2}, \tag{4.22}$$

$$\frac{\ddot{a}}{a} = -\frac{4\pi}{3}(\rho + 3p) + \frac{\Lambda}{3}. \tag{4.23}$$

This set of equations is often referred to as the Friedmann equations, they are the starting point for analysing cosmological solutions to the Einstein field equations and are the very basis of cosmology. In principle one could start the study of cosmology from these equations without much reference to General Relativity. While this approach is possible up to some point, more advanced subject in cosmology will again require the full machinery of differential geometry.

We recall that the energy–momentum tensor satisfies the conservation equation $\nabla_a T^a_b = 0$, and we must keep in mind that this equation is not independent of the field equations. It is implied by the field equations by virtue of the twice contracted Bianchi identities, see Theorem 1.6. One can see this explicitly by differentiating Eq. (4.22) with respect to time t, and elimination of the curvature parameter k from the Friedmann equations, which leads to

$$\dot{\rho} + 3\frac{\dot{a}}{a}(\rho + p) = 0. \tag{4.24}$$

One tends to refer to this equation as the cosmological energy–momentum conservation equation. This is equivalent to the equation $\nabla_a T^a_0 = 0$, the three equations $\nabla_a T^a_i = 0$ with $i = 1, 2, 3$ are identically satisfied.

Cosmological models based on the Friedmann–Lemaître–Robertson–Walker metric are described by two independent equations, two out of the three Eqs. (4.22)–(4.24). They contain three unknown functions, namely the scale factor $a(t)$, and the matter $\rho(t)$ and $p(t)$. Therefore, our equations are under-determined and we need some additional input to close this system. The most physical approach is to specify the matter content of the universe, this means we must chose an equation of state for the matter. In cosmology one typically considers a perfect fluid with linear equation of state $p = w\rho$ where w is the equation of state parameter. In classical mechanics w would correspond to the square root of the sound speed in this fluid.

When $w = 0$, the pressure vanishes and one speaks of a pressure-less perfect fluid which is often called a dust. In cosmology this is called a matter-dominated universe. For $w = 1/3$ one deals with radiation, recall that radiation also carries momentum which results in radiation having pressure. For standard matter the equation of state parameter satisfies $0 \leq w < 1$. The upper bound means we require the speed of sound to be less than the speed of light. When $w = 1$ one speaks of stiff matter, the sound speed would be the speed of light. Cosmology also deals with non-standard equations of state: for $-1 \leq w < -1/3$ one speaks of dark energy, $w = -1$ corresponds to the cosmological constant, and for $w < -1$ the fluid is called a phantom fluid.

For such a linear equation of state $p = w\rho$, the energy–momentum conservation equation (4.24) can be integrated using separation of variables which yields

$$\frac{\dot{\rho}}{\rho} = -3\frac{\dot{a}}{a}(1 + w), \quad \Rightarrow \quad \rho = \rho_0 a^{-3(1+w)}. \tag{4.25}$$

The constant of integration is denoted by ρ_0. We will interpret this result in the following section when discussing cosmological solutions in detail.

4.2. Cosmological Solutions

4.2.1. *Matter-dominated universe*

For a matter-dominated universe ($w = 0$) the conservation equation can be integrated and we find $\rho = \rho_0 a^{-3}$, see Eq. (4.25), which holds independently of the curvature parameter k. This is, of course, the expected result for matter when we recall the basic definition of density, namely mass or energy per volume. The scale factor a corresponds to a length scale and hence we can interpret a^3 as a volume. Since the cosmological Einstein field equations are two independent equations and we have already solved one of them, we are left with one more equation to be solved, namely Eq. (4.22). Upon substitution of the density in terms of the scale factor we have

$$\frac{\dot{a}^2}{a^2} = \frac{8\pi\rho_0}{3a^3} + \frac{\Lambda}{3} - \frac{k}{a^2}. \tag{4.26}$$

We can solve this equation for $\dot{a}$ and apply separation of variables which results in

$$t - t_0 = \int \left[\frac{8\pi\rho_0}{3a} + \frac{\Lambda}{3}a^2 - k\right]^{-\frac{1}{2}} da, \tag{4.27}$$

where t_0 is a constant of integration. The positive root will always be chosen so that the scale factor is positive. Our aim of finding the function $a(t)$ has been reduced to an integration problem. Unfortunately, integrals of this type are pretty hard and we cannot state an explicit solution in terms of elementary functions for all Λ and k. However, we can calculate this integral for some cases, thereby understanding the qualitative behaviour of the different types of possible solutions.

Let us begin by considering the simplest case, a matter-dominated universe with zero curvature parameter $k = 0$ and vanishing cosmological constant, $\Lambda = 0$. In this case, our integral simplifies considerably and we have

$$t - t_0 = \int \left[\frac{8\pi\rho_0}{3a}\right]^{-\frac{1}{2}} da = \frac{1}{\sqrt{8\pi\rho_0/3}}\frac{2}{3}a^{\frac{3}{2}}, \tag{4.28}$$

which we can now solve for the scale factor. The result is

$$a(t) = (6\pi\rho_0)^{\frac{1}{3}}(t - t_0)^{\frac{2}{3}}, \tag{4.29}$$

and we assume that the universe had zero volume at time t_0 so that $a(t_0) = 0$, moreover, we will set $t_0 = 0$, as the time when the universe began is arbitrary. The most important aspect of this solution is the scaling of $a(t)$ with respect to t, namely, $a(t) \propto t^{2/3}$. This implies that there exists a particle horizon. Moreover, $\rho(t) \propto t^{-2}$ so that the energy density diverges as $t \to 0$ which is called the big bang.

Next, we are interested in the matter-dominated universe with $k = 1$ in which case we need to integrate

$$t - t_0 = \int \left[\frac{8\pi\rho_0}{3a} - 1\right]^{-\frac{1}{2}} da = \int \frac{\sqrt{a}}{\sqrt{8\pi\rho_0/3 - a}}da. \tag{4.30}$$

We can evaluate this integral using the substitution $a = (8\pi\rho_0/3)\sin^2(u/2)$ so that we arrive at

$$t - t_0 = \int \frac{8\pi\rho_0}{3}\sin^2(u/2)du = \frac{4\pi\rho_0}{3}(u - \sin(u)). \tag{4.31}$$

It turns out that we cannot solve this equation explicitly for the scale factor $a(t)$ and have to accept a solution in parametric form which we can write as

$$a = \frac{4\pi\rho_0}{3}(1 - \cos u), \tag{4.32}$$

$$t = \frac{4\pi\rho_0}{3}(u - \sin u). \tag{4.33}$$

We set $t_0 = 0$ which means that $a(t = 0) = 0$. The curve described by these equations is in fact well known in mechanics, it is the standard

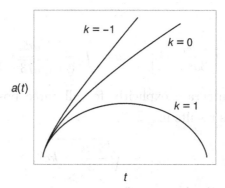

Fig. 4.1   Scale factor $a(t)$ for a matter-dominated universe without cosmological term, for the three different curvature parameters.

parametrisation of the cycloid. The cycloid is the curve traced by a point on the rim of a circle rolling along a straight line. A universe of this type will have a maximal size because of Eq. (4.32). The maximum is attained when $u = \pi$ which means we have

$$a_{\max} = \frac{8\pi\rho_0}{3}, \quad t_{\max} = \frac{4\pi^2\rho_0}{3}, \quad (4.34)$$

after which the universe will contract and decrease in size. At $t_{\text{end}} = 8\pi^2\rho_0/3$ the universe will have shrunk to $a = 0$. The final phase of this universe is called the big crunch opposed to the big bang. The scale factor for the matter-dominated universe is shown in Fig. 4.1.

4.2.2.   *Radiation-dominated universe*

Having discussed the matter-dominated solutions first, we now move on to the radiation-dominated solutions of the Einstein field equations. As before, we will restrict ourselves to $\Lambda = 0$ for now and deal with the cosmological term later. For a radiation-dominated universe, the energy–momentum conservation equation, Eq. (4.25) yields $\rho = \rho_0 a^{-4}$. The remaining Einstein field equation (4.22) now gives

$$t - t_0 = \int \left[\frac{8\pi\rho_0}{3a^2} - k \right]^{-\frac{1}{2}} da = \int \frac{a\,da}{\sqrt{8\pi\rho_0/3 - ka^2}}, \qquad (4.35)$$

which we can integrate explicitly for all three possible curvature parameters. This results in

$$t - t_0 = -\frac{1}{k} \sqrt{\frac{8\pi\rho_0}{3} - ka^2}, \qquad (4.36)$$

for $k = \pm 1$, while for $k = 0$ we find

$$t - t_0 = \left(\frac{8\pi\rho_0}{3} \right)^{-\frac{1}{2}} \frac{1}{2} a^2. \qquad (4.37)$$

Let us solve those equations for the scale factor

$$a(t) = \begin{cases} \sqrt{2}(8\pi\rho_0/3)\sqrt{t} & \text{if } k = 0, \\ \sqrt{(8\pi\rho_0/3) - (t - t_0)^2} & \text{if } k = +1, \\ \sqrt{(t - t_0)^2 - (8\pi\rho_0/3)} & \text{if } k = -1. \end{cases} \qquad (4.38)$$

For a spatially flat universe we have $a(t) \propto t^{1/2}$, $\rho(t) \propto t^{-2}$, and there exists a particle horizon. As in the matter-dominated case, the spatially-closed universe has a maximal size $a_{\max}$ and a finite age $t_{\max}$. The hyperbolic case is characterised by $a(t) \propto t$ for large t.

4.2.3.   *The Einstein static universe*

The Einstein static universe is a very particular solution of the Einstein field equations which is mainly of historical interest. Consider the field equations with $k = +1$, $p = 0$ and $\Lambda \neq 0$. Is it possible to find static solutions where $a = a_E$ and $\rho = \rho_E$ are constants? The field equation (4.22) gives

$$0 = \frac{8\pi\rho_E}{3} + \frac{\Lambda}{3} - \frac{1}{a_E^2} \quad \Rightarrow \quad a_E^{-1} = \sqrt{\frac{8\pi\rho_E}{3} + \frac{\Lambda}{3}}, \qquad (4.39)$$

which means that the 'size' of this universe is determined by the energy density and the cosmological constant. Moreover, from the other field equation (4.23) we find

$$0 = -\frac{4\pi}{3}\rho_E + \frac{\Lambda}{3} \quad \Rightarrow \quad \Lambda = 4\pi\rho_E. \tag{4.40}$$

Thus, the energy density and the cosmological constant cannot be chosen independently. Note that this solution only exists for positive Λ as energy densities cannot be negative. Without cosmological term this solution does not exist.

The most interesting aspect of this solution is that it is unstable with respect to small homogeneous perturbations of the scale factor and the energy density. This is a very nice calculation discussed in Exercise 4.8.

4.2.4. *De Sitter universe*

Let us consider the Einstein field equations with cosmological term $\Lambda \neq 0$, without any additional matter sources, this means $\rho = p = 0$. The one independent field equation is

$$\left(\frac{\dot{a}}{a}\right)^2 = \frac{\Lambda}{3} - \frac{k}{a^2}, \tag{4.41}$$

which we can solve using separation of variables. This gives

$$\sqrt{\frac{\Lambda}{3}}(t - t_0) = \int \frac{da}{\sqrt{a^2 - 3k/\Lambda}}, \tag{4.42}$$

which for $k = 0$ immediately results in

$$a(t) = a_0 e^{\sqrt{\Lambda/3}\,t}, \tag{4.43}$$

where a_0 is our new constant of integration. This solution is generally called the de Sitter solution, it is particularly interesting in the context of inflation. We note that the de Sitter universe cannot have started with a big bang at finite time in the past, unlike the matter and radiation-dominated models.

The general solution to Eq. (4.42) can be written in the form

$$a(t) = a_0 \cosh(\sqrt{\Lambda/3}t) + a_0\sqrt{1 - \frac{3k}{\Lambda a_0^2}} \sinh(\sqrt{\Lambda/3}t), \qquad (4.44)$$

which for $k = 0$ reduces to (4.43). As before, the constant of integration is chosen such that $a(0) = a_0$.

For the de Sitter solution with $k = 0$ the resulting line element is

$$ds^2 = -dt^2 + a_0^2 e^{2\sqrt{\Lambda/3}t} \left(dx^2 + dy^2 + dz^2\right), \qquad (4.45)$$

and has some very interesting properties which are not at all obvious at first sight. First, it appears that this metric is non-static, however, we can show that it is in fact static. Let us consider a time translation $t \mapsto \tilde{t} + T$ where T is some constant, which means the metric becomes

$$ds^2 \mapsto ds^2 = -d\tilde{t}^2 + a_0^2 e^{2\sqrt{\Lambda/3}(\tilde{t}+T)} \left(dx^2 + dy^2 + dz^2\right). \qquad (4.46)$$

Since $\exp(2\sqrt{\Lambda/3}(\tilde{t} + T)) = \exp(2\sqrt{\Lambda/3}\tilde{t})\exp(2\sqrt{\Lambda/3}T)$ we note that the spatial part of the metric is rescaled. Introducing new spatial coordinates $\tilde{x} = \exp(\sqrt{\Lambda/3}T)x$ and similarly for the other two coordinates, we find that the de Sitter metric, after time translation and rescaling of the spatial coordinates, is given by

$$ds^2 \mapsto ds^2 = -d\tilde{t}^2 + a_0^2 e^{2\sqrt{\Lambda/3}\tilde{t}} \left(d\tilde{x}^2 + d\tilde{y}^2 + d\tilde{z}^2\right), \qquad (4.47)$$

which is identical to Eq. (4.45). Therefore the de Sitter metric is form-invariant under time translation and hence static. This is slightly surprising, it turns out that de Sitter space has a larger symmetry group than the other cosmological metrics we encountered. De Sitter space has some other interesting mathematical properties, for instance it can be mapped to part of a 5D hyperboloid. Our choice of coordinates only covers part of the entire manifold.

4.3. Physical Cosmology

Our next aim is to connect the underlying geometry with quantities that can be determined by observations. The Friedmann–Lemaître–Robertson–Walker metric contains only one function, the scale factor $a(t)$, and the curvature parameter k. Recalling that one function already contains an infinite number of degrees of freedom, it is clear that we cannot determine the exact functional form of the scale factor by observations. This is intuitively clear, no experiment can access the form of $a(t)$ for $t > t_{\text{today}}$. However, a viable approach would be to make a series expansion of $a(t)$ around t_{today} with the aim of determining the first few coefficients in this series by observations. Having said that, we could of course solve the field equations for specific forms of matter and determine the functional form of $a(t)$ for all times and compare this with cosmological observations. It turns out that cosmology follows both these approaches.

4.3.1. *Cosmological parameters*

Let us begin here with the aforementioned series expansion of $a(t)$ at the time $t_0 = t_{\text{today}}$. Conventionally, quantities with subscript 0 refer to their values today, where today really refers to the current cosmological time, and we will write $a_0 = a(t_0)$. Therefore we have

$$a(t) = a_0 + \dot{a}_0(t - t_0) + \ddot{a}_0 \frac{(t - t_0)^2}{2!} + \cdots . \qquad (4.48)$$

The current size of the universe a_0 is a matter of convention, often this quantity is simply set to one, however, we will keep a_0 arbitrary for now. Since we cannot determine this number, we will write our series as follows:

$$a(t) = a_0 \left[1 + \frac{\dot{a}_0}{a_0}(t - t_0) + \frac{\ddot{a}_0}{a_0} \frac{(t - t_0)^2}{2!} + \cdots \right]. \qquad (4.49)$$

The quantity $\dot{a}_0/a_0$ corresponds to an expansion or contraction velocity of the universe, depending on the sign or $\dot{a}_0$. We define the Hubble function

$$H = \frac{\dot{a}}{a},\tag{4.50}$$

which measures the rate of expansion. We note that this quantity is invariant under rescaling of $a(t)$ by some constant, $a(t) \mapsto \lambda a(t)$ which follows from its definition. Today's value of H is denoted by H_0 and is called the Hubble parameter. It is conventional to use units of km/s/Mpc to give the measured value of H_0. This corresponds to units of $1/s$ which is consistent with the units in Eq. (4.50).

Since the quantity H_0 has units of inverse time, the quantity $x = (t - t_0)H_0$ is dimensionless. Our next step is to re-write Eq. (4.49) using this quantity which leads to

$$a(t) = a_0 \left[1 + x - q_0 \frac{x^2}{2!} + \cdots\right],\tag{4.51}$$

where we introduced the new dimensionless parameter $q_0 = -\ddot{a}/(aH^2)$. The number q_0 is called the deceleration parameter, the minus sign in the definition and also the name are historically motivated. It measures the rate of change of the acceleration of the universe. In principle one can continue with the series (4.51) to include higher-order terms in x, and introduce further dimensionless constants. For the present purpose it is sufficient to discuss the Hubble parameter and the deceleration parameter. The definitions to these two parameters are based on the Friedmann–Lemaître–Robertson–Walker metric and are independent of the Einstein field equations, one could speak of kinematic quantities.

On the other hand, let us now recall the Einstein field equation (4.22) using the Hubble parameter

$$H^2 = \frac{8\pi\rho}{3} + \frac{\Lambda}{3} - \frac{k}{a^2}.\tag{4.52}$$

Division by H^2 on both sides of this equation yields an equation which naturally contains only dimensionless quantities

$$1 = \frac{8\pi\rho}{3H^2} + \frac{\Lambda}{3H^2} - \frac{k}{a^2 H^2}. \tag{4.53}$$

This leads to the definition of the density parameters in cosmology

$$\Omega = \frac{8\pi\rho}{3H^2} = \frac{\rho}{\rho_{\text{crit}}}, \quad \Omega_\Lambda = \frac{\Lambda}{3H^2}. \tag{4.54}$$

Here ρ_{crit} is the so-called critical energy density which is given by $\rho_{\text{crit}} = 3H^2/(8\pi)$ and we should note that this is a function of time in general. In models with matter and radiation one distinguishes the different density parameters with additional subscripts.

Next, we introduce the total energy density $\Omega_{\text{total}} = \Omega + \Omega_\Lambda$ which allows us to write the first field equation in the very neat form

$$\Omega_{\text{total}} - 1 = \frac{k}{a^2 H^2}. \tag{4.55}$$

Despite its simplicity, this is one of the most important equations in cosmology. In cosmological models where $k \neq 0$, the total density Ω_{total} is a function of time. However, if $k = 0$ then $\Omega_{\text{total}} = 1$ at all times. Hence, in spatially flat cosmological models the total density is determined from the very beginning. We can also read this equation the other way around. If we were able to accurately measure or estimate the matter content of the universe, we could determine the curvature of the constant time hypersurfaces. This is quite amazing really! Note that we are only interested in the sign of k since its numerical values can always be rescaled. What we are witnessing here is the very essence of General Relativity, namely that geometry and matter are inextricably linked via the Einstein field equations.

4.3.2.   *Redshift*

Electromagnetic signals in General Relativity and Cosmology are described by null geodesic. In particular we are interested in radial geodesics where θ and ϕ remain constant. The Lagrangian describing such geodesics follows from Eq. (4.7) and is given by

$$L = -\dot{t}^2 + a(t)^2 \dot{\chi}^2. \tag{4.56}$$

Since we are considering null geodesics, $L = 0$ and we immediately find

$$\chi = \pm \int_{t_0}^{t_1} \frac{dt}{a(t)}, \tag{4.57}$$

where the sign and the limits of this integral depend on some conventions. We choose coordinates so that the observer is located at the origin, and as the signal was emitted before it was received, $t_1 < t_0$. Since we consider signals travelling towards the observer at the origin, we must choose the negative sign.

In order to clarify the notation we will now denote $t_e = t_1$, and $t_0 = t_{obs}$ where the subscripts stand for emitter and observer, respectively. Hence, the final form of our geodesic equation is

$$\chi = \int_{t_e}^{t_{obs}} \frac{dt}{a(t)}. \tag{4.58}$$

The quantity χ is called the comoving distance in cosmology. We already noted that for a $k = 0$ universe this is equal to the metric distance, which is simply $d_{metric} = \chi$. It is no coincidence that this is identical to the newly introduced conformal time coordinate in Eq. (4.12).

Let us now assume that a first signal is emitted at time t_e and that a second signal is sent at time $t_e + \delta t_e$. The emitter is assumed to be at rest, more precisely comoving, by which we mean that it follows the expansion of the universe only and has no intrinsic motion relative to the observer. The comoving distance is unchanged between the two signals so that

$$\int_{t_e+\delta t_e}^{t_{obs}+\delta t_{obs}} \frac{dt}{a(t)} - \int_{t_e}^{t_{obs}} \frac{dt}{a(t)} = 0. \tag{4.59}$$

We split the first integral into three parts by changing the limit. Instead of integrating from $t_e + \delta t_e$ to $t_{obs} + \delta t_{obs}$, we integrate from t_e to t_{obs}, then from t_e to $t_{obs} + \delta t_{obs}$, and then need to subtract the integral from t_e to $t_e + \delta t_e$. This second integral will cancel the last term in Eq. (4.59) and we arrive at

$$\int_{t_{obs}}^{t_{obs}+\delta t_{obs}} \frac{dt}{a(t)} - \int_{t_e}^{t_e+\delta t_e} \frac{dt}{a(t)} = 0. \tag{4.60}$$

Next we assume that $\delta t_{obs} \ll 1$ and $\delta t_e \ll 1$, so that we can apply the fundamental theorem of calculus which yields

$$\frac{\delta t_{obs}}{a(t_{obs})} - \frac{\delta t_e}{a(t_e)} = 0, \tag{4.61}$$

plus corrections of higher order in the small parameters. We can interpret δt_{obs} and δt_e as the time intervals between any two maxima of the signal's wave when observed or emitted. Hence we can write $\delta t_{obs} = \lambda_{obs}$ and $\delta t_e = \lambda_e$, so that Eq. (4.61) simply becomes $\lambda_{obs}/\lambda_e = a(t_{obs})/a(t_e)$. The standard definition of the redshift is the relative difference between the observed and emitted wavelengths, this quantity is usually denoted by z. Applied to our cosmological setting we have

$$z = \frac{\lambda_{obs} - \lambda_e}{\lambda_e} = \frac{\lambda_{obs}}{\lambda_e} - 1 = \frac{a(t_{obs})}{a(t_e)} - 1. \tag{4.62}$$

Therefore, the gravitational redshift due to the universe's expansion is determined entirely by the scale factor at the time of emission and time of observation. This should not be too surprising as our cosmological model is based on homogeneity and isotropy and only contains one unknown function in the metric. Our final result is the simple formula

$$1 + z = \frac{a(t_{\text{obs}})}{a(t_{\text{e}})}. \tag{4.63}$$

Let us briefly return to Eq. (4.51) and consider this in linear order in x, this means we can write

$$\frac{a(t)}{a_0} = 1 + (t - t_0)H_0. \tag{4.64}$$

Using our redshift formula, and multiplying the entire equation by the speed of light c we obtain

$$z\,c = (t - t_0)c\,H_0. \tag{4.65}$$

The quantity $z\,c$ has units of velocity and is referred to as the redshift velocity, it can be interpreted as the recessional velocity of the galaxy emitting the electromagnetic signal. The quantity $(t-t_0)c$ has units of length and corresponds to the distance between the emitting galaxy and the observer. This is nothing but Hubble's law

$$v = D H_0, \tag{4.66}$$

first observed in 1929. This was the first observational evidence to support an expanding universe, hence supporting a dynamic and not a static universe. Note that Hubble's law and our derivation of it are independent of the Einstein field equations, in the sense that we studied geodesics on a manifold with metric (4.7) and considered a series expansion of the scale factor.

4.3.3. *Distances in cosmology*

There are different notions of distances used in cosmology, most of which will reduce to the Euclidean distance for nearby objects. It is common to express these distances using the redshift as this is one of the directly observable quantities in cosmology.

Let us begin by introducing the luminosity distance d_L. Consider an object with absolute luminosity L, this is the total power radiated by that object. Light and other radiation expand from the object in spherical wavefronts. Thus, at Euclidean distance d_L, the apparent luminosity l, that is the power received per unit area by an observer, is simply $L = 4\pi d_L^2 l$. This total power L spreads over the sphere of radius d_L. We can use this relation as our definition of the luminosity distance

$$d_L = \sqrt{\frac{L}{4\pi l}}, \tag{4.67}$$

which must be placed into the cosmological setting we are interested in. We assume the emitting galaxy to have redshift z, at time t_e and with metric distance $d_{\text{metric}} = \Sigma(\chi)$. From Eq. (4.7) we know that the effective radius of the sphere is $a(t_{\text{obs}})\Sigma(\chi_e)$. Therefore, the electromagnetic signal will have spread over the area

$$A = 4\pi a(t_{\text{obs}})^2 \Sigma(\chi_e)^2. \tag{4.68}$$

We denote by L_{received} the total power received at the effective radius $a(t_{\text{obs}})\Sigma(\chi_e)$, so that we can write the apparent luminosity as $l = L_{\text{received}}/A$. As the absolute luminosity has units of energy per unit time we can also write $L_{\text{received}} = \Delta E_{\text{obs}}/\Delta t_{\text{obs}}$. Rewriting Eq. (4.62) using frequencies results in $\nu_e/\nu_{\text{obs}} = a(t_{\text{obs}})/a(t_e)$ which implies that the emitted and observed energies will obey the same relation. Likewise, we have $\Delta t_{\text{obs}}/\Delta t_e = a(t_{\text{obs}})/a(t_e)$. Hence we arrive at

$$L_{\text{received}} = \frac{\Delta E_{\text{obs}}}{\Delta t_{\text{obs}}} = \frac{1}{(1+z)^2}L, \tag{4.69}$$

which means that the apparent luminosity is given by

$$l = \frac{L_{\text{received}}}{A} = \frac{L}{(1+z)^2 4\pi a(t_{\text{obs}})^2 \Sigma(\chi_e)^2}. \tag{4.70}$$

Inserting this latter relation into our luminosity distance definition (4.67) gives

$$d_L = (1+z)a(t_{\text{obs}})\Sigma(\chi_e), \tag{4.71}$$

which is the equation we wanted to derive. We note that $\Sigma(\chi_e)$ is the difficult part of this equation. Determining this function requires knowledge of the comoving distance χ at the emitter (see Eq. (4.57)) which in turn needs the functional form of $a(t)$ which is a solution of the cosmological Einstein field equations. It is at this point where everything comes together, we will discuss some concrete examples and also state a generic formula useful in this context.

Let us introduce the angular diameter distance first. Let us begin with a galaxy of diameter D and angular diameter δ. If this galaxy has Euclidean distance d_A from the observer, we have

$$\frac{D}{2d_A} = \tan\frac{\delta}{2} \approx \frac{\delta}{2}, \tag{4.72}$$

for small angular diameters. As before, we use this relation as the definition of the angular diameter distance

$$d_A = \frac{D}{\delta}. \tag{4.73}$$

Next, we must reconsider this in the setting of an expanding universe. This is much simpler than in the previous case. When measuring the distance across the diameter of the galaxy we have $t = t_e$, $\chi = \chi_e$ and we can always align our coordinates such that $\phi = 0$. Hence, the diameter D is given by

$$D = \int_{-\delta/2}^{\delta/2} \sqrt{(g_{22})_e}\, d\theta = a(t_e)\Sigma(\chi_e) \int_{-\delta/2}^{\delta/2} d\theta = a(t_e)\Sigma(\chi_e)\delta, \tag{4.74}$$

where $(g_{22})_e$ means that this metric component is evaluated at the emitter. Therefore, we find for the angular diameter distance

$$d_A = a(t_e)\Sigma(\chi_e). \tag{4.75}$$

We can relate the scale factor at the time of emission to that of the time of observation using the redshift Eq. (4.62) which yields the final form

$$d_A = \frac{1}{1+z} a(t_{\text{obs}}) \Sigma(\chi_e).$$ (4.76)

Comparison of the luminosity distance d_L with the angular diameter distance d_A shows that they differ by a factor of $(1+z)^2$, this means we have the simple relationship

$$d_L = (1+z)^2 d_A.$$ (4.77)

In the following section we will show how to derive explicit formulae for the distance redshift relationships for specific cosmological models.

Some other distances can be defined in cosmology, see for instance Hogg (1999), which also contains references to some of the literature.

4.3.4. *Distance redshift relationships*

After defining the luminosity distance d_L and the angular diameter distance d_A, we are now in a position to derive their explicit forms for different cosmological models.

Let us begin with considering a matter-dominated universe, $w = 0$, without cosmological term, $\Lambda = 0$, assuming the spatially flat case, $k = 0$. We know from Sec. 4.2.1 that $a(t) = a_0 t^{2/3}$, we also have the Hubble parameter $H = 2/(3t)$, and recall the definition of the redshift, Eq. (4.63) which reads $1 + z = a(t_{\text{obs}})/a(t_e)$. Our aim is to calculate $\Sigma(\chi_e)$ explicitly.

Beginning with Eq. (4.8) we have

$$\Sigma(\chi_e) = \chi_e = \int_{t_e}^{t_{\text{obs}}} \frac{dt}{a(t)} = \frac{1}{a_0} \int_{t_e}^{t_{\text{obs}}} \frac{dt}{t^{\frac{2}{3}}} = \frac{3}{a_0} (t_{\text{obs}}^{\frac{1}{3}} - t_e^{\frac{1}{3}})$$

$$= \frac{3}{a_0} t_{\text{obs}}^{\frac{1}{3}} \left(1 - \frac{t_e^{\frac{1}{3}}}{t_{\text{obs}}^{\frac{1}{3}}}\right).$$ (4.78)

Since we conduct observations today, we have $H_0 = 2/(3t_{\rm obs})$ and therefore

$$d_L = (1 + z)a(t_{\rm obs})\Sigma(\chi_e) = (1 + z)(a_0 t_{\rm obs}^{2/3})\Sigma(\chi_e)$$

$$= \frac{2}{H_0}(1 + z)\left(1 - \frac{t_e^{\frac{1}{3}}}{t_{\rm obs}^{\frac{1}{3}}}\right) = \frac{2}{H_0}(1 + z)\left(1 - (1 + z)^{-\frac{1}{2}}\right). \quad (4.79)$$

In the ultimate step we used the redshift definition together with the explicit form of $a(t)$. Moving the factor $(1 + z)$ into the bracket leads to our final result

$$d_L = \frac{2}{H_0}\left((1 + z) - \sqrt{1 + z}\right), \quad (4.80)$$

for the luminosity distance. This also determines the angular diameter distance

$$d_A = \frac{d_L}{(1 + z)^2} = \frac{2}{H_0}\left((1 + z)^{-1} - (1 + z)^{-\frac{3}{2}}\right). \quad (4.81)$$

Let us emphasise here that both results (4.80) and (4.81) are only valid for the specific model we chose.

It is instructive to consider the small z approximation of d_L using $\sqrt{1 + z} = 1 + z/2 + \cdots$ which results in

$$d_L = \frac{2}{H_0}\left((1 + z) - (1 + z/2 + \cdots)\right) = \frac{z}{H_0} + \cdots, \quad (4.82)$$

where the dots stand for some higher-order corrections. This is precisely Hubble's law (4.66) and at this point it is a consequence of the Einstein field equations since we worked with an explicit solution.

Similar calculations can be performed for different cosmological models, however, it might not always be possible to arrive at nice explicit solutions for these quantities. Hence, let us next study these distances in a more general setting.

In Eq. (4.58), we can change the integration variable from time to redshift by writing

$$\frac{dt}{dz} = \frac{dt}{da}\frac{da}{dz} = \frac{1}{\dot{a}}\frac{da}{dz} = \frac{1}{aH}\frac{da}{dz}. \qquad (4.83)$$

On the other hand, we have $a(z) = a(t_{\text{obs}})/(1+z)$ so that

$$\frac{da}{dz} = -\frac{a(t_{\text{obs}})}{(1+z)^2}, \qquad (4.84)$$

which can be combined to

$$\frac{dt}{dz} = -\frac{1}{aH}\frac{a(t_{\text{obs}})}{(1+z)^2} = -\frac{1}{(1+z)H}. \qquad (4.85)$$

We must note that the Hubble parameter is now viewed as a function of the redshift $H(z)$. We introduce a new function $H(z) = H_0 E(z)$ with $E(0) = 1$ so that we can rewrite Eq. (4.58) as follows:

$$\chi_e = \int_z^0 \frac{1}{a(\bar{z})}\frac{dt}{d\bar{z}}d\bar{z} = \frac{1}{a_{\text{obs}}H_0}\int_0^z \frac{1}{E(\bar{z})}d\bar{z}. \qquad (4.86)$$

Now, we must determine the function $E(z)$ from the field equations. Using Eq. (4.52) we can write

$$E(z)^2 = \frac{8\pi\rho_{\text{total}}}{3H_0^2} + \frac{\Lambda}{3H_0^2} - \frac{k}{a^2 H_0^2}, \qquad (4.87)$$

which has similarities with the cosmological density parameters. For the sake of completeness we also introduce the parameter $\Omega_k = -k/(a_{\text{obs}}^2 H_0^2)$. Recall the critical density evaluated today $\rho_{\text{crit0}} = 3H_0^2/(8\pi)$. Moreover we assume $\rho_{\text{total}} = \rho_m + \rho_{\text{rad}}$ with their corresponding conservation equations expressed using redshifts. Then

$$\frac{\rho_m}{\rho_{\text{crit0}}} = \Omega_{m0}(1+z)^3, \qquad \frac{\rho_{\text{rad}}}{\rho_{\text{crit0}}} = \Omega_{\text{rad0}}(1+z)^4. \qquad (4.88)$$

Substituting everything back into Eq. (4.87) gives us the explicit expression

$$E(z)^2 = \Omega_{m0}(1+z)^3 + \Omega_{\text{rad0}}(1+z)^4 + \Omega_{\Lambda 0} + \Omega_k(1+z)^2, \qquad (4.89)$$

which only depends on the redshift and the values of the parameters today, this means at the time of the observations. Since the explicit form of $E(z)$ in general contains a square root of a polynomial in z, it is clear that the integral in Eq. (4.86) might be quite difficult to compute.

We can also integrate Eq. (4.85) directly with respect to the redshift, to find the difference between the age of the universe at observation t_{obs} and emission t_{e} which gives

$$t_{\text{obs}} - t_{\text{e}} = - \int_z^0 \frac{d\bar{z}}{(1+\bar{z})H} = \frac{1}{H_0} \int_0^z \frac{d\bar{z}}{(1+\bar{z})E(\bar{z})}. \qquad (4.90)$$

This quantity is sometimes called the lookback time. By assuming that the time of emission of a signal was approximately at the beginning of the universe, we can set $t_{\text{e}} = 0$. Since we assume that $a(t) \to 0$ as $t \to 0$, we also have $z \to \infty$ as we approach the beginning. Hence, we can determine the age of the universe for a specific cosmological model by

$$t_{\text{universe}} = \frac{1}{H_0} \int_0^\infty \frac{dz}{(1+z)E(z)}. \qquad (4.91)$$

Example 4.1 (Flat radiation-dominated universe and refocussing). Let us compute the luminosity distance in a spatially flat, $k = 0$, radiation-dominated universe, $w = 1/3$, without cosmological term, $\Lambda = 0$. The function $E(z)$ is given by Eq. (4.89) by setting $\Omega_{\text{m}0} = \Omega_{\Lambda 0} = \Omega_k = 0$ so that $E(z)^2 = \Omega_{\text{rad}0}(1+z)^4$. Since radiation is the only matter source, we also have $\Omega_{\text{rad}0} = 1$. Therefore, the comoving distance χ is given by

$$\chi = \frac{1}{a_0 H_0} \int_0^z \frac{1}{(1+\bar{z})^2} d\bar{z} = \frac{1}{a_0 H_0} \left(1 - \frac{1}{1+z} \right). \qquad (4.92)$$

Since $\Sigma(\chi_{\text{e}}) = \chi_{\text{e}}$ in a spatially flat universe, we find

$$d_L = (1+z)a(t_{\text{obs}})\chi_{\text{e}} = \frac{1}{H_0}z. \qquad (4.93)$$

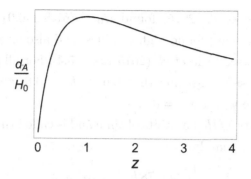

Fig. 4.2   Angular diameter distance redshift relation.

which looks again like Hubble's law, but valid for all redshifts. Consequently, the angular diameter distance is

$$d_A = \frac{d_L}{(1+z)^2} = \frac{1}{H_0}\frac{z}{(1+z)^2}. \tag{4.94}$$

The shape of d_A as the redshift changes is quite intriguing, see Fig. 4.2. There exits a maximum value for the angular diameter distance at $z = 1$. For larger values of redshift, the angular diameter distance decreases which seems rather counter intuitive. Moreover, any given angular diameter distance corresponds to two different redshifts. This effect is called refocusing in cosmology.

4.3.5.   *The universe today*

After discussing various aspects of cosmology, let us now put this into context with our universe and current observations which determine the various physical parameters discussed. One of the most important quantities is today's value of the Hubble parameter H_0. The physical units of H_0 are inverse time, however, it is usually written as

$$H_0 = 100h\frac{\text{km}}{\text{s\,Mpc}}, \tag{4.95}$$

where h is a dimensionless constant. Measuring H_0 is notoriously difficult and there are many uncertainties about its value. Planck

2015 data Ade *et al.* (2015) found $H_0 = (67.8 \pm 0.9)\text{kms}^{-1}\text{Mpc}^{-1}$, however, others have found higher values, and also lower values have been reported, see [Ade *et al.* (2015), Sec. 5.4]. For all practical purposes we can safely assume that $0.6 < h < 0.8$, and for explicit calculations we will use $h = 0.7$.

The quantity $1/H_0$ has units of time and is called the Hubble time which is approximately

$$\frac{1}{H_0} = 9.78 \, h^{-1} \times 10^9 \, \text{year}. \tag{4.96}$$

However, the actual age of the universe is larger with the Planck data suggesting $t_{\text{universe}} = 13.8 \times 10^9$ year.

When multiplying the Hubble time with the speed of light c we get a quantity with units of length. This so-called Hubble length is approximately

$$\frac{c}{H_0} = 2998 \, h^{-1} \, \text{Mpc}. \tag{4.97}$$

Next, we need to know the matter content of the universe which is also linked to its spatial geometry. Let us begin with the curvature parameter for which observations suggest $|\Omega_k| < 0.005$, this means that the curvature of the constant time slices is very close to zero. If we take this as evidence to set $k = 0$, then by Eq. (4.55) we will have $\Omega_{\text{total}} = 1$ at all times. However, we will see in Sec. 4.3.6 that k close to zero yields some problems in cosmology. There is strong evidence for the presence of a cosmological term Λ with Ω_Λ making up slightly less than 70% of the contents of the universe, so that the total matter content is roughly $\Omega_m = 0.3$.

The matter density Ω_m accounts for all matter which interacts gravitationally, it includes radiation and neutrinos, for instance, but also any form of luminous and non-luminous matter. Ordinary matter in the universe is referred to as baryons because protons and neutrons account for most of its density. The baryon density is denoted by Ω_b and its value is approximately $\Omega_b = 0.05$ which means that ordinary matter only makes up about 5% of the contents of the universe we observe. In turn this implies that the remaining 25% must be in form of some non-luminous matter which interacts gravitationally and has only very weak interactions with other matter. This form of matter is called dark matter and is subject to substantial research. There exists a large number of different dark matter models which are inspired by a diverse range of physical theories. In analogy to dark matter, one speaks of dark energy when theories are explored which mimic the cosmological constant. Dark matter and dark energy taken together make up approximately 95% of the matter content of the universe and it is somewhat fair to say that we do not understand either of them from a theoretical point of view.

We should also mention the critical density of the universe observed today which is approximately $(\rho_{\mathrm{crit}})_0 = 1.88\,h^2 \times 10^{29}\,\mathrm{g\,cm^{-3}}$. Let us recall that the proton mass is $m_p = 1.76 \times 10^{-24}\,\mathrm{g}$, then we can conclude that the critical density observed today corresponds to about one proton per cubic meter.

In cosmology, it is possible to define the temperature of the universe at any given time. We will not derive these relations rigorously but only state the main results which are needed for some of the subsequent discussions.

The Stefan–Boltzmann law of power radiated from a black body reads $\rho \propto T^4$. Next we recall that in a radiation-dominated universe $\rho \propto a^{-4}$ from which we conclude the relationship $T \propto a^{-1}$. This allows us to relate the temperature to the scale factor and by virtue of Eq. (4.63) to the redshift z. We are led to

$$\frac{T(t_{\mathrm{e}})}{T(t_{\mathrm{obs}})} = \frac{a(t_{\mathrm{obs}})}{a(t_{\mathrm{e}})} = 1 + z, \tag{4.98}$$

where $T(t_{\mathrm{obs}})$ generally corresponds to the temperature we observe today and $T(t_{\mathrm{e}})$ corresponds to the temperature of the universe at some time in the past. The idea here is that we are interested in the temperature of the universe when a signal was emitted at t_e which we can observe today.

In a matter-dominated universe we are dealing with non-relativistic particles for which we assume the Maxwell–Boltzmann distribution that states $\rho \propto T^{3/2}$. Moreover, in a matter-dominated universe we have $\rho \propto a^{-3}$ which implies the relationship $T \propto a^{-2}$. Hence we arrive at

$$\frac{T(t_e)}{T(t_{obs})} = \frac{a(t_{obs})^2}{a(t_e)^2} = (1+z)^2. \tag{4.99}$$

This immediately implies that matter cools down much faster than radiation in an expanding universe, we also see that a smaller universe (matter or radiation dominated) in the past was much hotter.

At a temperature of about $T(t_{rec}) = 3 \times 10^3$ K photons decouple from matter which means that hydrogen can form once the universe has cooled sufficiently. In cosmology this is called decoupling or recombination, note the misnomer as these objects were never combined previously. We observe an almost uniform cosmological microwave background radiation which corresponds to a temperature of roughly $T_{(rad)}(t_{obs}) = 2.73$ K. If we interpret this radiation as the cooled photon gas at decoupling, the moment the photons could travel freely, we can estimate the redshift of recombination using

$$z_{rec} = \frac{T(t_{rec})}{T_{(rad)}(t_{obs})} - 1 \approx 1100. \tag{4.100}$$

This tells us that the universe was about 1,100 times smaller at the time of decoupling.

We can also estimate the temperature of the matter part of the universe. At the time of recombination matter and radiation had the same temperature so that we have

$$T_{(matter)}(t_{today}) = \frac{T(t_{rec})}{(1+z_{rec})^2} \approx 2.5 \times 10^{-3} \text{K}, \tag{4.101}$$

which is three orders of magnitude cooler than the radiation.

4.3.6. *Shortcomings in cosmology*

(a) The flatness problem

Let us begin by recalling Eq. (4.55) which relates the total energy content of the universe to its spatial curvature

$$\Omega_{\text{total}} - 1 = \frac{k}{a^2 H^2}. \qquad (4.102)$$

We already noted that in a spatially flat universe $\Omega_{\text{total}} = 1$ at all times. When $k \neq 0$, the density parameter evolves with time, and this is determined by aH. For a radiation-dominated universe we found $a \propto t^{1/2}$ and $H = 1/(2t)$, so that $aH \propto t^{-1/2}$, and for a flat matter dominated universe $a \propto t^{2/3}$ and $H = 2/(3t)$, so that $aH \propto t^{-1/3}$. Let us now consider a universe with a small non-vanishing curvature parameter. We assume that its evolution is close to that of the flat case which implies $\Omega_{\text{total}} - 1 \propto t$ in the radiation-dominated case and $\Omega_{\text{total}} - 1 \propto t^{2/3}$ in the matter-dominated case. The point here is that the quantity $\Omega_{\text{total}} - 1$ is an increasing function with time. Since we observe $\Omega_{\text{total}}(t_{\text{today}})$ to be of order unity, we must conclude that Ω_{total} was very close to one at earlier cosmological times. An estimate based on nucleosynthesis calculations requires $|\Omega_{\text{total}}(t_{\text{nucleo}}) - 1| \leq 10^{-16}$ at which point the universe was about one second old. At times earlier than this, the density parameter of universe would have to be even closer to one. The flatness problem simply states that such finely tuned initial conditions for the universe appear to be unlikely. In other words, let us prescribe some random initial conditions for the universe at some very early time. It turns out that in almost all cases the universe will evolve very differently to the universe we observe, and that our present universe is in fact highly improbable.

(b) Horizon problem

In Sec. 4.1.4 we discussed the presence of particle horizons. Let us assume that the universe began at $t = 0$ with $a(0) = 0$. Then the present distance of an object which emitted light at the beginning is given by $a_0\chi(t_e)$ so that for a flat universe we have

$$d_{\text{hor}}(t) = a(t_{\text{today}}) \int_0^t \frac{dt'}{a(t')}. \tag{4.103}$$

Using the explicit forms of $a(t)$ for a radiation dominated and matter-dominated universe, respectively, we find the following particle horizons

$$d_{\text{hor}}(t) = 3t = \frac{2}{H(t)} \quad \text{matter-dominated}, \tag{4.104}$$

$$d_{\text{hor}}(t) = 2t = \frac{1}{H(t)} \quad \text{radiation-dominated}. \tag{4.105}$$

Currently we are in a matter-dominated universe so that today's particle horizon is $d_{\text{hor}}(t_{\text{today}}) = 2/H_0$, which is twice the Hubble length that was mentioned in Eq. (4.97).

Given that the age of the universe is about $t_{\text{universe}} = 13.8 \times 10^9$ yr, we can find the time of recombination by using Eq. (4.98) which gives

$$\frac{a(t_{\text{today}})}{a(t_{\text{rec}})} = \left(\frac{t_{\text{today}}}{t_{\text{rec}}} \right)^{\frac{2}{3}} = 1 + z_{\text{rec}} \approx 10^3, \tag{4.106}$$

which implies approximately $t_{\text{rec}} = 4.4 \times 10^5$ yr. Prior to the decoupling of the photons, the universe was radiation-dominated and hence the particle horizon at recombination was $d_{\text{hor}}(t_{\text{rec}}) = 2t_{\text{rec}}$. Reinserting the speed of light c into the distance to the particle horizon gives

$$d_{\text{hor}}(t_{\text{today}}) = 3ct_{\text{today}} = \frac{2c}{H_0} \approx 5996\, h^{-1}\, \text{Mpc}, \tag{4.107}$$

$$d_{\text{hor}}(t_{\text{rec}}) = 2ct_{\text{rec}} \approx 0.27\, \text{Mpc}. \tag{4.108}$$

At first sight, the horizon appears to be moving with a velocity faster than the speed of light which would contradict special relativity. The reason for this is simply that the universe is expanding while the photon is travelling. Locally photons move with the speed of light.

When the photons decoupled at t_{rec} the distance over which causal interactions could have occurred is approximately 0.27 Mpc.

However, the cosmic microwave radiation we observe today is fairly homogeneous and isotropic over much larger scales, namely those corresponding to our particle horizon scale. This fact is usually referred to as the horizon problem. This means regions across the sky separated by about one to two degrees ($\theta \approx (z_{\rm rec})^{-1/2} \approx 0.03 \approx 1.7°$) cannot have interacted before recombination and hence there is no physical reason why these regions should look similar. This means that these causally disconnected regions should have evolved independently. However, we observe that these regions are statistically indistinguishable. Homogeneity and isotropy of the cosmic microwave background radiation therefore must have been encoded into the universe at early times of its evolution. As mentioned in the above, it is very unlikely for a universe with randomly chosen initial conditions to have this property.

There are some additional issues with respect to particles (or topological defects) forbidden to be present by observations which are predicted by some theoretical models applicable at very high temperatures. These are referred to as unwanted relics but we will not discuss them in more detail.

4.4. Inflation

The original motivation for inflationary cosmology came from the desire to solve these shortcomings in cosmology. However, the most important aspect of inflation is that it can generate the initial irregularities in the early universe which led to the formation of larger structures like galaxies. Inflation models also predict tensor perturbations which would yield gravitational waves.

In simple words, inflation is an early time epoch of the cosmological evolution where the universe's acceleration is positive. It turns out that this single requirement suffices to resolve the flatness problem and the horizon problem in a very elegant way.

4.4.1.   *Accelerated expansion*

We can formally define inflation by the condition

$$\ddot{a} > 0. \tag{4.109}$$

Note that neither the scale factor of the matter-dominated universe nor the radiation-dominated universe satisfy this condition, in both cases the acceleration is a decreasing function of time. Interestingly, the de Sitter solution, discussed in Sec. 4.2.4, has positive acceleration and hence is relevant in the context of inflation. Next, we wish to reformulate the inflation condition (4.109) using the Hubble parameter as this will provide us with more physical definitions.

From the definition of $H = \dot{a}/a$ we have $\dot{H} = \ddot{a}/a - H^2$ so that our inflation condition can also be written as

$$0 < \frac{\ddot{a}}{a} = \dot{H} + H^2, \quad \Leftrightarrow \quad -\frac{\dot{H}}{H^2} < 1, \tag{4.110}$$

which is always satisfied for positive $\dot{H}$. Another useful formulation of the inflation condition begins with $\dot{a} = Ha$ and again differentiation with respect to time. We have

$$\frac{d}{dt}\left(\frac{H^{-1}}{a}\right) = \frac{d}{dt}\left(\frac{1}{\dot{a}}\right) = -\frac{\ddot{a}}{\dot{a}^2} < 0, \tag{4.111}$$

because of $\ddot{a} > 0$ and $\dot{a}^2 > 0$. The interesting part in this way of defining inflation is the quantity H^{-1}/a. Recall that H^{-1} is the Hubble length (in units where $c = 1$). The quantity H^{-1}/a is the comoving Hubble length and this length scale determines whether two regions of spacetime can communicate now, in other words it sets the size of the observable universe. This means that during an epoch of accelerated expansion the observable universe becomes smaller!

Before discussing the implications of this particular aspect further, we can give a third definition of inflation based on the matter content of the universe. Recall Eq. (4.23) without cosmological constant

$$\frac{\ddot{a}}{a} = -\frac{4\pi}{3}(\rho + 3p). \tag{4.112}$$

Hence, the condition $\ddot{a} > 0$ immediately implies that the matter must satisfy the condition

$$\rho + 3p < 0. \tag{4.113}$$

Since the energy density ρ is always assumed to be positive, we conclude that inflation requires matter to have negative pressure which violates the so called strong energy condition. For a linear equation of state $p = w\rho$, condition (4.113) becomes $w < -1/3$, a value which one also encounters when discussing the stability of the Einstein static universe in Exercise 4.9.

In order to understand how inflation solves the shortcomings of standard cosmology discussed in Sec. 4.3.6 we will primarily need to focus on the inflation condition (4.111) which states that during inflation the comoving Hubble length is decreasing. This allows for the following possibility: the length scale of the particle horizon could be much larger than the comoving Hubble length today. However, if the size of the observable universe rapidly decreased during a period of accelerated expansion, then those regions which are separated by the particle horizon today might have been in causal contact in the past. Therefore, inflation provides us with a simple solution to the horizon problem. Moreover, the flatness problem stemmed from the fact that the function $1/(a^2 H^2)$ is increasing for a radiation or matter-dominated universe. Inflation is defined precisely by the opposite condition, $1/(a^2 H^2)$ is a decreasing function, and hence $|\Omega - 1|$ is driven towards zero rather than away from it, thereby solving the flatness problem.

4.4.2. *Scalar fields in cosmology*

The simplest matter field which satisfies the inflation condition $\rho + 3p < 0$ is a scalar field which we shall call ϕ. Despite scalar fields being used in cosmology for about three decades, the existence of scalar fields like the Higgs boson in the standard model of particle physics has only been confirmed by the Large Hadron Collider at CERN in 2012 and 2013. Independent of the particle nature of the field driving inflation in the early universe, the scalar field provides an ideal model to study a concrete inflation model which shows good agreement with observations. Moreover, the scalar field is mathematically quite simple and so allows the explicit computation of many effects which would be much harder for a more complicated field.

In Sec. 2.4, we introduced the variational approach to General Relativity which also contained a definition of the energy–momentum tensor by the matter Lagrangian, see Eq. (2.127). The scalar field action is given by

$$S_\phi = \int \mathcal{L}_\phi \sqrt{-g}\, d^4x = \int \left[-\frac{1}{2} g^{ab} \nabla_a \phi \nabla_b \phi - V(\phi) \right] \sqrt{-g}\, d^4x,$$
(4.114)

where $\phi = \phi(x^i)$ is the scalar function depending on all the coordinates, and $V(\phi)$ is a potential which depends only on the scalar field. Different scalar field inflation models are characterised by their potentials. The energy–momentum tensor is given by

$$T_{ab} = \nabla_a \phi \nabla_b \phi + g_{ab} \mathcal{L}_\phi.$$
(4.115)

Since ϕ is a scalar field, we could write partial derivatives instead of covariant derivatives, however, the use of the covariant derivative makes it clear that $\nabla_a \phi$ is a tensor. Since we are interested in a cosmological scalar field, we assume ϕ to be a function of time only, and consider the Friedmann–Lemaître–Robertson–Walker metric. In this case the energy density and pressure are

$$\rho_\phi = \frac{1}{2}\dot{\phi}^2 + V(\phi), \tag{4.116}$$

$$p_\phi = \frac{1}{2}\dot{\phi}^2 - V(\phi). \tag{4.117}$$

The scalar field has its own equation of motion which we can derive by considering the variations of Eq. (4.114) with respect to ϕ. Alternatively, we recall that any cosmological matter satisfies the conservation equation $\dot{\rho} + 3H(\rho + p) = 0$ and hence we can compute the conservation equation of the scalar field directly from ρ_ϕ and p_ϕ. We have

$$\dot{\rho}_\phi = \ddot{\phi}\dot{\phi} + \frac{dV}{d\phi}\dot{\phi}, \quad \rho + p_\phi = \dot{\phi}^2, \tag{4.118}$$

so that the conservation equation yields

$$\dot{\phi}\left(\ddot{\phi} + 3H\dot{\phi} + \frac{dV}{d\phi}\right) = 0. \tag{4.119}$$

This can be satisfied by either a constant scalar field which would then be equivalent to a cosmological constant, or provided that $\dot{\phi} \neq 0$, the scalar field has to satisfy the so-called Klein–Gordon equation

$$\ddot{\phi} + 3H\dot{\phi} + \frac{dV}{d\phi} = 0. \tag{4.120}$$

The scalar field driving inflation is generally referred to as the inflaton.

Using Eqs. (4.116) and (4.117) we can define an effective equation of state for the scalar field

$$w_\phi = \frac{p_\phi}{\rho_\phi} = \frac{\frac{1}{2}\dot{\phi}^2 - V(\phi)}{\frac{1}{2}\dot{\phi}^2 + V(\phi)}, \tag{4.121}$$

which is not a constant but a function of time. This equation also provides us with support that a scalar field can give rise to a period of accelerated expansion. If the kinetic energy $\dot{\phi}^2/2$ is much smaller than the potential energy $V(\phi)$, the effective equation of state will give $w_\phi \approx -1$. This corresponds to a cosmological constant which we know provides accelerated expansion.

When studying inflation models it is common to work with the reduced Planck mass M_{Pl} instead of the gravitational coupling constant κ. The reduced Planck mass is defined by

$$M_{\text{Pl}} = \sqrt{\frac{\hbar c}{8\pi G}}, \quad \text{or} \quad M_{\text{Pl}}^{-2} = \frac{8\pi G}{\hbar c}, \tag{4.122}$$

which when working in natural units where $G = c = \hbar = 1$, simply becomes $M_{\text{Pl}}^{-2} = 8\pi$. In these units the scalar field has dimensions of mass which are useful in the context of particle physics. Using the Planck mass, the cosmological field equations, with scalar fields as matter sources, are given by

$$H^2 = \frac{1}{3M_{\text{Pl}}^2}\left(\frac{1}{2}\dot{\phi}^2 + V(\phi)\right), \tag{4.123}$$

$$-\frac{\ddot{a}}{a} = \frac{1}{3M_{\text{Pl}}^2}\left(\dot{\phi}^2 - V(\phi)\right), \tag{4.124}$$

$$\ddot{\phi} + 3H\dot{\phi} + \frac{dV}{d\phi} = 0. \tag{4.125}$$

The first definition of inflation was simply the requirement $\ddot{a} > 0$, therefore Eq. (4.124) implies that this condition is satisfied provided that $\dot{\phi}^2 < V(\phi)$. Physically speaking this condition means that we need a model with small kinetic energy relative to the potential energy, we can think of slow motions here. For instance a sufficiently flat potential should eventually satisfy this condition as the particle loses kinetic energy. This leads us directly to slow-roll inflation which is a very useful approximation method for inflation models.

4.4.3. *Slow-roll inflation*

When introducing the effective equation of state (4.121) we noted that we can achieve accelerated expansion provided that the kinetic energy is much smaller than the potential energy. This also follows directly from Eq. (4.124). We will make this more formal and introduce the slow-roll approximation as follows: we assume $\dot{\phi}^2 \ll V(\phi)$ and $\ddot{\phi} \ll 1$. The second condition, slow accelerations, is natural when one thinks about slow motions. However, small velocities do not imply small accelerations. One could have a situation where the velocity is always small but changes very rapidly. For example, the velocity $v(t) = 0.1 \times \sin(100t)$ is small, then the acceleration $a(t) = \dot{v}(t) = 10 \times \cos(100t)$ is clearly large compared to the velocity.

In this approximation, our field equations (4.123) and (4.125) simplify to

$$H^2 \simeq \frac{1}{3M_{\mathrm{Pl}}^2} V(\phi), \qquad (4.126)$$

$$3H\dot{\phi} \simeq -\frac{dV}{d\phi}. \qquad (4.127)$$

The symbol $\simeq$ means that both sides are equal within the slow-roll approximation. Our original equations were of second order and nonlinear in $\dot{\phi}$ and ϕ. In the slow-roll approximation the equations are first order and can, in principle, be solved using separation of variables.

Two convenient parameters are the so-called slow-roll parameters ϵ and η which are defined by

$$\epsilon = \frac{M_{\mathrm{Pl}}^2}{2} \left(\frac{V'}{V}\right)^2, \quad \eta = M_{\mathrm{Pl}}^2 \frac{V''}{V}, \qquad (4.128)$$

where the prime denotes differentiation with respect to the scalar field. We say the slow-roll approximation is valid if $\epsilon \ll 1$ and $\eta \ll 1$, and we should take note that these quantities only depend on the potential V. In particular the parameter ϵ can be motivated nicely. We recall our inflation condition (4.120) which read $-\dot{H}/H^2 < 1$.

Differentiating Eq. (4.126) with respect to time gives

$$2H\dot{H} \simeq \frac{1}{3M_{\text{Pl}}^2}V'\dot{\phi} \simeq \frac{1}{3M_{\text{Pl}}^2}V'\left(-\frac{V'}{3H}\right), \qquad (4.129)$$

where for the second part we used Eq. (4.127). Dividing the latter equation by $-2H^3$ yields

$$-\frac{\dot{H}}{H^2} \simeq \frac{M_{\text{Pl}}^2}{2}\frac{(V')^2}{9M_{\text{Pl}}^4 H^4} \simeq \frac{M_{\text{Pl}}^2}{2}\left(\frac{V'}{V}\right)^2 = \epsilon. \qquad (4.130)$$

Therefore, our inflation condition (4.120) is satisfied provided that $\epsilon < 1$. One could also use relation (4.130) as the defining equation for the slow-roll parameter via $\epsilon = -\dot{H}/H^2$. It turns out that this quantity only depends on the shape of the potential in this approximation.

During inflation the scale of the universe typically increases by about 10^{26} which is a very large number, so we are primarily interested in the exponent. In order to quantify the amount of inflation one considers the ratio of the scale factor at the end of inflation to its values at some time before the end of inflation. We define the number of e-foldings by

$$\mathcal{N}(t) = \log\frac{a(t_{\text{end}})}{a(t)}, \qquad (4.131)$$

which seems slightly counter-intuitive. We note that $\mathcal{N}(t_{\text{end}}) = 0$, therefore the number of e-foldings at some time t measures the amount of inflation that still has to occur after that time. Typically, the number of e-foldings is 50–70 in order to solve the flatness and horizon problems satisfactorily.

Within the slow-roll approximation we can explicitly calculate this quantity without ever needing to solve the cosmological field equations. This is one of the great features of this approximation, we can check whether a specific potential $V(\phi)$ can give rise to inflation. We can define the end of inflation, for instance using $\epsilon = 1$ or $\eta = 1$. Lastly, we can compute the amount of inflation that has taken place by computing the number of e-foldings. To compute $\mathcal{N}(t)$, we begin with recalling that $H = \dot{a}/a$ so that

$$\mathcal{N}(t) = \log \frac{a(t_{\text{end}})}{a(t)} = \int_t^{t_{\text{end}}} H(t')dt'. \qquad (4.132)$$

However, we can now exploit the slow-roll field equations. From Eq. (4.127) we find $dt = -(3H/V')d\phi$ while Eq. (4.127) allows us to express H in term of the potential. This yields

$$\mathcal{N}(t) \simeq - \int_{\phi(t)}^{\phi_{\text{end}}} \frac{3H^2}{V'}d\phi \simeq \frac{1}{M_{\text{Pl}}^2} \int_{\phi_{\text{end}}}^{\phi(t)} \frac{V}{V'}\,d\phi, \qquad (4.133)$$

which again only depends on the potential and does not involve the scale factor or solutions to field equations. We will finish with two examples to see the various features of the slow-roll approximation worked out explicitly for two potentials.

Example 4.2 (Massive potential $V(\phi)=m^2\phi^2/2$). Consider the potential $V(\phi) = m^2\phi^2/2$ where m would correspond to the mass of the scalar field. We have $V'(\phi) = m^2\phi$, and $V''(\phi) = m^2$ so that

$$\frac{V'}{V} = \frac{2}{\phi}, \quad \frac{V''(\phi)}{V(\phi)} = \frac{2}{\phi^2}. \qquad (4.134)$$

We can now compute the slow-roll parameters using their definitions and find

$$\epsilon = M_{\text{Pl}}^2 \frac{2}{\phi^2}, \quad \eta = M_{\text{Pl}}^2 \frac{2}{\phi^2}. \qquad (4.135)$$

For this particular potential both parameters are identical. Our condition for inflation $\epsilon \ll 1$ implies that $\phi \gg \sqrt{2}M_{\text{Pl}}$. If we define the end of inflation by the condition $\epsilon = 1$, we find that $\phi_{\text{end}} = \sqrt{2}M_{\text{Pl}}$.

Next, let us compute the total number of e-foldings for this potential. We have

$$\mathcal{N} \simeq \frac{1}{M_{\text{Pl}}^2} \int_{\phi_{\text{end}}}^{\phi_{\text{i}}} \frac{V}{V'} d\phi = \frac{1}{M_{\text{Pl}}^2} \int_{\phi_{\text{end}}}^{\phi_{\text{i}}} \frac{1}{2} \phi \, d\phi$$

$$= \frac{1}{M_{\text{Pl}}^2} \frac{1}{4} \phi^2 \Big|_{\phi_{\text{end}}}^{\phi_{\text{i}}} = \frac{1}{M_{\text{Pl}}^2} \frac{1}{4} \left(\phi_{\text{i}}^2 - \phi_{\text{end}}^2 \right) = \left(\frac{\phi_{\text{i}}}{2M_{\text{Pl}}} \right)^2 - \frac{1}{2}. \quad (4.136)$$

This means that choosing $\phi_{\text{i}} = 15 M_{\text{Pl}}$ yields a little over 60 e-foldings. Let us emphasise again that we did not solve the field equations so far.

So, let us finish this example by solving the slow-roll approximated field equations. Equation (4.126) can be solved for the Hubble parameter to get

$$H \simeq \frac{1}{\sqrt{6}} \frac{m}{M_{\text{Pl}}} \phi, \quad (4.137)$$

which we can substitute into the second equation (4.127) to give

$$\sqrt{\frac{3}{2}} \frac{m}{M_{\text{Pl}}} \phi \dot{\phi} \simeq -m^2 \phi. \quad (4.138)$$

We can cancel a factor of $m\phi$ on both sides and integrate directly with respect to time, the result of which is

$$\phi(t) = -\sqrt{\frac{2}{3}} M_{\text{Pl}} m t + C, \quad (4.139)$$

where C is a constant of integration. We already know that we require $\phi_{\text{i}} = 15 M_{\text{Pl}}$ and $\phi_{\text{end}} = \sqrt{2} M_{\text{Pl}}$ for this model to give the right number of e-foldings. Setting the time t_{i}, when inflation began, to zero $t_{\text{i}} = 0$, we can fix the constant of integration to give $C = 15 M_{\text{Pl}}$ and find the solution

$$\phi(t) = M_{\text{Pl}} \left(15 - \sqrt{\frac{2}{3}} m t \right). \quad (4.140)$$

Since we also know the field value at the end inflation, we can compute t_{end} from $\phi(t_{\text{end}}) = \sqrt{2}M_{\text{Pl}}$ which yields

$$t_{\text{end}} = \frac{\sqrt{3}}{m}\left(\frac{15}{\sqrt{2}} - 1\right). \qquad (4.141)$$

We are not done yet as we still need to determine the scale factor $a(t)$. Since $H(t) = \dot{a}/a = d\log(a)/dt$ we can directly integrate Eq. (4.137) using the explicit solution for the scalar field. This gives

$$\log(a(t)) = -\frac{m^2}{6}t^2 + 5\sqrt{\frac{3}{2}}mt + \log(D), \qquad (4.142)$$

where D is another constant of integration. Since we assumed that inflation began at $t_{\text{i}} = 0$, we have $a(t_{\text{i}}) = D$ so that D corresponds to the scale of the universe at that point in time, we denote this by a_{i}. Hence our final scale factor is given by

$$a(t) = a_{\text{i}} \exp\left(-\frac{m^2}{6}t^2 + 5\sqrt{\frac{3}{2}}mt\right). \qquad (4.143)$$

As a final consistency check we will verify that $\ddot{a}/a > 0$ during inflation. Differentiating the result (4.143) twice with respect to time and dividing by $a(t)$ yields

$$\frac{\ddot{a}}{a} = 18m^2\left(2m^2t^2 - 30\sqrt{6}mt + 669\right). \qquad (4.144)$$

Clearly, at t_i the acceleration is positive while at t_{end} we find $\ddot{a}(t_{\text{end}}) = 0$ in agreement with our conceptual framework of inflation.

Example 4.3 (Power-law inflation). Power-law inflation is a solution to the field equations based on the following potential:

$$V(\phi) = V_0 \exp\left(-\sqrt{\frac{2}{p}}\frac{\phi}{M_{\text{Pl}}}\right), \qquad (4.145)$$

where V_0 and p are positive constants. We begin with computing the slow-roll parameters using $V' = -\left(\sqrt{2/p}/M_{\text{Pl}}\right)V$ and

$V'' = (2/(pM_{\mathrm{Pl}}^2))V$ so that we find

$$\epsilon = \frac{1}{p}, \quad \eta = \frac{2}{p}. \tag{4.146}$$

Therefore, the inflation condition is satisfied provided that $p > 1$. However, unlike in the previous model, there is no natural end to inflation since p is a constant. Therefore, if the inflation condition is satisfied at one point in time, it will be satisfied at all times. For such an inflation model one would need to introduce an additional mechanism to stop inflation. However, solving the slow-roll approximated field equations gives a particularly nice solution which we will now derive.

From the first field equation (4.126) we get

$$H \simeq \frac{1}{M_{\mathrm{Pl}}} \sqrt{\frac{V_0}{3}} \exp\left(-\sqrt{\frac{1}{2p}} \frac{\phi}{M_{\mathrm{Pl}}}\right). \tag{4.147}$$

Substitution into the second field equation (4.127) results in

$$\sqrt{3V_0} \exp\left(-\sqrt{\frac{1}{2p}} \frac{\phi}{M_{\mathrm{Pl}}}\right) \dot{\phi} \simeq \sqrt{\frac{2}{p}} V_0 \exp\left(-\sqrt{\frac{2}{p}} \frac{\phi}{M_{\mathrm{Pl}}}\right), \tag{4.148}$$

which we can write as follows:

$$\exp\left(\sqrt{\frac{1}{2p}} \frac{\phi}{M_{\mathrm{Pl}}}\right) d\phi = \sqrt{\frac{2}{3p}} \sqrt{V_0} dt. \tag{4.149}$$

Now both sides can be integrated and we arrive at

$$M_{\mathrm{Pl}} \sqrt{2p} \exp\left(\sqrt{\frac{1}{2p}} \frac{\phi}{M_{\mathrm{Pl}}}\right) = \sqrt{\frac{2}{3p}} \sqrt{V_0} \, (t - t_0). \tag{4.150}$$

Here t_0 is some constant of integration. Let us substitute this result back into Eq. (4.147) in order to find $H(t)$. It turns out that almost all terms cancel nicely and we are left with

$$H = \frac{p}{t - t_0} \quad \Rightarrow \quad a(t) = a_{\mathrm{i}}(t - t_0)^p, \tag{4.151}$$

which justifies the name power-law inflation of this model. Recall that inflation never ends so it would be meaningless to compute the number of e-foldings without a mechanism to stop it.

4.5. Further Reading

The theory of inflation was originally introduced to resolve the shortcomings of big bang cosmology. More importantly, inflation provides us with a mechanism to create the initial density perturbations (quantum fluctuations) of the early universe that seed the growth of larger gravitating structures. These will eventually evolve into the objects we see in today's night sky. The theory of inflation is very successful in that it is able to provide us with the appropriate initial conditions compatible with the universe we observe today. Connecting the fluctuations of the inflaton in the very early universe with the Einstein field equations and extracting observable quantities out of this model is at the heart of modern cosmology. Three books which would be a natural continuation along these lines are by Dodelson (2003), Weinberg (2008) and Liddle (2015). A very different book is by Rowan-Robinson (2004) which emphasises the empirical evidence and observational data which underlies cosmology.

In order to study how small fluctuations of densities affect the evolution of the universe and how such fluctuations evolve within an expanding universe, one needs to study cosmological perturbation theory. This is very similar to the weak gravity approach we considered in Sec. 2.3. The main difference is that instead of considering small perturbations about Minkowski space, one is interested in small perturbations about the Friedmann–Lemaître–Robertson–Walker metric. As in Sec. 2.3.2 one can choose a particular gauge in which the perturbed metric takes a particularly nice form. For instance, when studying scalar perturbations with $k = 0$

one can write

$$ds^2 = -(1 + 2\Phi)dt^2 + a^2(t)(1 - 2\Psi)(dx^2 + dy^2 + dz^2), \quad (4.152)$$

where Φ and Ψ are functions of all the coordinates and are considered to be small. Using this metric one can compute the Einstein field equations in linear order in the perturbations, along the lines of the weak field limit. Cosmological perturbation theory is an important tool of modern cosmology, see for instance the book by Gorbunov and Rubakov (2011).

Most topics in modern cosmology focus on the Friedmann–Lemaître–Robertson–Walker metric and study physics in this setting. However, one could also study cosmological models which are homogeneous but anisotropic, this is very interesting in its own right and some of the shortcomings in cosmology are less of an issue in this framework. The interested reader is referred to the book by Ryan and Shepley (1975) which discusses such models in detail.

The recommended literature is by no means complete, it simply reflects the author's suggestions for further reading.

4.6. Exercises

Classical and modern cosmology

Exercise 4.1. Consider an infinitely large and eternal universe which is static, and contains an infinite number of stars. Such a universe should imply the night sky to be very bright, instead of dark. This is known as Olbers paradox. Propose various explanations to solve this paradox.

Exercise 4.2. Construct an argument based on thermodynamics to deduct that the universe cannot be infinitely old (heat death paradox or Clausius paradox).

Exercise 4.3. Calculate the non-vanishing Christoffel symbol components of the Friedmann–Lemaître–Robertson–Walker metric given by Eq. (4.6).

Exercise 4.4. Following on from the previous exercise, compute the Ricci tensor components of (4.6) and verify that the Ricci scalar is given by Eq. (4.16).

Exercise 4.5 (hard). Verify that the volume of the 3-sphere $\mathbb{S}^3$ is given by $V(\mathbb{S}^3) = 2\pi^2$, and also show that $V(\mathbb{S}^2) = 4\pi$ and $V(\mathbb{S}^4) = 8\pi^2/3$. One has to be slightly careful with the notion of volume at this point as the unit sphere $\mathbb{S}^2$ embedded in $\mathbb{E}^3$ is usually said to have area 4π. However, when $\mathbb{S}^2$ is viewed as a manifold, we can visualise this manifold as the surface area of a sphere. The volume of this manifold corresponds to its surface, thus the confusion.

The interested reader might wish to push this exercise further and show that the volume of the general n-dimensional sphere $\mathbb{S}^n$ is given by $2\pi^{n/2}/\Gamma(n/2)$ where Γ is the gamma function, $\Gamma(1/2) = \pi$, and $\Gamma(n+1) = n\Gamma(n)$ (no complete solution given for this final part).

Cosmological solutions

Exercise 4.6. Solve the cosmological field equations assuming a matter-dominated universe $w = 0$, without cosmological term $\Lambda = 0$ and with spatial negative curvature $k = -1$.

Exercise 4.7. Generalise the Einstein static universe by including pressure.

Exercise 4.8. Show that the Einstein static universe, Sec. 4.2.3, is unstable with respect to small homogeneous perturbation, $a_E \rightarrow a_E + \delta a(t)$ and $\rho_E \rightarrow \rho_E + \delta\rho(t)$.

Exercise 4.9. Include pressure into the previous calculation, assume the equation of state $p = w\rho$, and show that one can have stable perturbation provided $-1 < w < -1/3$.

Exercise 4.10. Show that the de Sitter solution for $k/\Lambda > 0$ can also be written in the form

$$a(t) = \sqrt{\frac{3k}{\Lambda}} \cosh\left(\sqrt{\Lambda/3}(t - t_0)\right).$$

Hint: Start with Eq. (4.42) and use the appropriate substitution.

Physical cosmology

Exercise 4.11. Use Eq. (4.91) to find the age of the universe assuming it is spatially flat and matter-dominated with $\Lambda = 0$. Where have we encountered a similar result?

Exercise 4.12. Find the luminosity distance in a spatially flat and matter dominated universe using Eq. (4.86).

Exercise 4.13. In the radiation-dominated universe with $\Lambda = 0$ and $k \neq 0$, show that

$$\Omega = \frac{\Omega_0(1 + z)^2}{1 + \Omega_0 z(A + z)}, \tag{4.153}$$

where $\Omega_0 = \Omega(t_0)$ is the present value of the density parameter. Determine the value of the constant A. Using this result find a relationship between the Hubble parameter H and the quantities H_0, Ω_0, z.

Exercise 4.14. Consider the spatially spherical universe, filled with radiation, without cosmological constant. Consider a light signal travelling along the azimuth so that $\chi = \theta = \pi/2$. Show that a light ray emitted at the 'big bang' will have travelled halfway around the universe by the time of the 'big crunch'.

Exercise 4.15. Consider the spatially spherical universe, filled with matter $(w = 0)$, without cosmological constant. Consider a light signal travelling along the azimuth so that $\chi = \theta = \pi/2$. Show that a light ray emitted at the 'big bang' will have travelled all the way around the universe by the time of the 'big crunch'.

Exercise 4.16. The laboratory wavelength of the so-called hydrogen$-\alpha$ line is 656.3 nm. An elliptic galaxy has the visible angular size $\theta = 4'$. The hydrogen$-\alpha$ line in its spectrum has the wavelength 657.0 nm. Find the distance to the galaxy and its proper size, using $H_0 = 60$ km/(s Mpc).

Inflation

Exercise 4.17. Show that the cosmological equations are equivalent to the autonomous nonlinear system

$$\dot{H} = -F(H, \phi), \quad \dot{\phi} = \sqrt{\frac{1}{4\pi}F(H, \phi)}. \qquad (4.154)$$

Find the function $F(H, \phi)$ explicitly. Assuming a constant potential $V = V_0 > 0$, integrate the cosmological field equations to find $H(t)$ and $a(t)$. Show that for large t, $a(t) \propto \exp(\mu(t - t_0))$ and find μ.

Exercise 4.18. Derive Eq. (4.115) using the variational approach and definition (2.127) .

Exercise 4.19 (takes time). Derive the cosmological Klein–Gordon equation (4.120) using the variational approach and metric (4.7) with $k = 0$.

Exercise 4.20. Show the field equations in slow-roll approximation imply

$$3\ddot{\phi} \simeq (\eta - \epsilon)V',$$

and argue for the condition $\eta \ll 1$.

5

Solutions to Exercises

5.1. Solutions: Differential Geometry

The concept of a vector

Exercise 1.1. The area of the parallelogram spanned by the vectors b, c is $A = |b \times c| = bc \sin \phi$, and the normal to this parallelogram is $(b \times c)/|b \times c|$. We now project a along this normal to find the height of the parallelepiped

$$h = \left| a \cdot \frac{b \times c}{|b \times c|} \right| = \frac{|a \cdot (b \times c)|}{|b \times c|}. \tag{5.1}$$

To find the volume, we multiply the area A by the height h and arrive at

$$V = Ah = |b \times c| \frac{|a \cdot (b \times c)|}{|b \times c|} = |a \cdot (b \times c)|. \tag{5.2}$$

Exercise 1.2. Following on from the previous exercise, we see that the volume of the parallelepiped spanned a, b, c is given by $V = |a \cdot (b \times c)|$. We can also find the volume by first computing the area of the parallelogram spanned by the vectors a, b, and then project c along the corresponding normal which would result in the volume $c \cdot (a \times b)$. Lastly, we can begin with the parallelogram spanned by a, c, however, we have to use the normal such that it points in the same direction as b. Hence the normal would be $c \times a$ and the volume would be $b \cdot (c \times a)$. Alternatively, we could write out all vectors in terms of basis vectors $a = a_1 i + a_2 j + a_3 k$ and compute the triple vector products explicitly using Eqs. (1.7) and (1.8).

Exercise 1.3. We have $e_i \cdot e^j = \delta_i^j$ and $e_i \times e_j = \varepsilon_{ijk} e^k$ and hence

$$a \cdot (b \times c) = a^s e_s \cdot (\varepsilon_{ijk} b^i c^j e^k) = a^s \varepsilon_{ijk} b^i c^j (e_s \cdot e^k)$$

$$= \varepsilon_{ijk} a^s b^i c^j \delta_s^k = \varepsilon_{ijk} b^i c^j a^k = \varepsilon_{kij} a^k b^i c^j. \qquad (5.3)$$

This also explains directly the symmetry properties of the scalar triple product.

Exercise 1.4. We write out the determinant explicitly

$$\varepsilon_{ijk} \varepsilon^{lmn} = \delta_i^l (\delta_j^m \delta_k^n - \delta_k^m \delta_j^n) - \delta_i^m (\delta_j^l \delta_k^n - \delta_k^l \delta_j^n) + \delta_i^n (\delta_j^l \delta_k^m - \delta_k^l \delta_j^m). \qquad (5.4)$$

Let us sum over the indices k and n which gives

$$\varepsilon_{ijk} \varepsilon^{lmk} = \delta_i^l (\delta_j^m \delta_k^k - \delta_k^m \delta_j^k) - \delta_i^m (\delta_j^l \delta_k^k - \delta_k^l \delta_j^k) + \delta_i^k (\delta_j^l \delta_k^m - \delta_k^l \delta_j^m)$$

$$= \delta_i^l (3\delta_j^m - \delta_j^m) - \delta_i^m (3\delta_j^l - \delta_j^l) + (\delta_j^l \delta_i^m - \delta_i^l \delta_j^m)$$

$$= \delta_i^l \delta_j^m - \delta_i^m \delta_j^l. \qquad (5.5)$$

Let us continue with this result and sum also over j and m which gives $\varepsilon_{ijk} \varepsilon^{ljk} = \delta_i^l \delta_j^j - \delta_i^j \delta_j^l = 3\delta_i^l - \delta_i^l = 2\delta_i^l$. Lastly, we arrive at $\varepsilon_{ijk} \varepsilon^{ijk} = 2\delta_i^i = 6$, where we repeatedly used $\delta_i^i = 3$ in three dimensions.

Exercise 1.5. We start with

$$b \times c = \begin{pmatrix} b_1 \\ b_2 \\ b_3 \end{pmatrix} \times \begin{pmatrix} c_1 \\ c_2 \\ c_3 \end{pmatrix} = \begin{pmatrix} b_2 c_3 - b_3 c_2 \\ b_3 c_1 - b_1 c_3 \\ b_1 c_2 - b_2 c_1 \end{pmatrix}, \qquad (5.6)$$

and then

$$a \times (b \times c) = \begin{pmatrix} a_1 \\ a_2 \\ a_3 \end{pmatrix} \times \begin{pmatrix} b_2 c_3 - b_3 c_2 \\ b_3 c_1 - b_1 c_3 \\ b_1 c_2 - b_2 c_1 \end{pmatrix}$$

$$= \begin{pmatrix} a_2 (b_1 c_2 - b_2 c_1) - a_3 (b_3 c_1 - b_1 c_3) \\ a_3 (b_2 c_3 - b_3 c_2) - a_1 (b_1 c_2 - b_2 c_1) \\ a_1 (b_3 c_1 - b_1 c_3) - a_2 (b_2 c_3 - b_3 c_2) \end{pmatrix}. \qquad (5.7)$$

Next we compute the right-hand side of the identity $(a \cdot c)b - (a \cdot b)c$ which is

$$\begin{pmatrix} b_1(a_1c_1 + a_2c_2 + a_3c_3) \\ b_2(a_1c_1 + a_2c_2 + a_3c_3) \\ b_3(a_1c_1 + a_2c_2 + a_3c_3) \end{pmatrix} - \begin{pmatrix} c_1(a_1b_1 + a_2b_2 + a_3b_3) \\ c_2(a_1b_1 + a_2b_2 + a_3b_3) \\ c_3(a_1b_1 + a_2b_2 + a_3b_3) \end{pmatrix}$$

$$= \begin{pmatrix} b_1(a_2c_2 + a_3c_3) - c_1(a_2b_2 + a_3b_3) \\ b_2(a_1c_1 + a_3c_3) - c_2(a_1b_1 + a_3b_3) \\ b_3(a_1c_1 + a_2c_2) - c_3(a_1b_1 + a_2b_2) \end{pmatrix}. \quad (5.8)$$

This agrees with the left-hand side we computed. Starting with the left-hand side again gives

$$a \times (b \times c) = \varepsilon_{lmn} a_l (\varepsilon_{ijm} b_i c_j) e_n = -\varepsilon_{mln} \varepsilon_{mij} a_l b_i c_j e_n$$

$$= -(\delta_{li}\delta_{nj} - \delta_{lj}\delta_{in}) a_l b_i c_j e_n$$

$$= \delta_{lj}\delta_{in} a_l b_i c_j e_n - \delta_{li}\delta_{nj} a_l b_i c_j e_n$$

$$= a_j c_j b_n e_n - a_i b_i c_n e_n = (a \cdot c)b - (a \cdot b)c, \quad (5.9)$$

and we are done using the index notation.

Exercise 1.6. We have $e_i \times e_j = \varepsilon_{ijs} e^s$, therefore $(e_i \times e_j) \cdot e_k = \varepsilon_{ijs} e^s \cdot e_k = \varepsilon_{ijs} \delta_k^s = \varepsilon_{ijk}$ which is what we wanted to show.

Exercise 1.7. The first two identities follow from the symmetry properties of ε_{ijk}. We have $\operatorname{div} \operatorname{curl} A = \varepsilon^{ijk} \partial_k (\partial_i A_j) = 0$, where we note that $\partial_k \partial_i$ is symmetric in the indices ki which ε^{ijk} is skew-symmetric. Likewise $\operatorname{curl} \operatorname{grad} f = \varepsilon^{ijk} \partial_i (\partial_j f) e_k = 0$. For the third identity we write

$$\operatorname{div}(A \times B) = (A \times B)^k{}_{,k} = (\varepsilon^{ijk} A_i B_j)_{,k}$$

$$= (\varepsilon^{ijk} \partial_k A_i) B_j + (\varepsilon^{ijk} \partial_k B_j) A_i$$

$$= \operatorname{curl} A \cdot B - \operatorname{curl} A \cdot B. \quad (5.10)$$

The final one is proved as follows:

$$\mathrm{div}(f\boldsymbol{A}) = \partial_k(fA^k) = A^k\partial_k f + f\partial_k A^k$$
$$= \boldsymbol{A}\cdot\mathrm{grad}\, f + f\,\mathrm{div}\,\boldsymbol{A}. \tag{5.11}$$

Manifolds and tensors

Exercise 1.8. Recall the transformation property of the partial derivative

$$\partial'_a A'^b = \frac{\partial X^d}{\partial X'^a}\frac{\partial\partial X'^b}{\partial X^d\partial X'^c}A^c + \frac{\partial X^d}{\partial X'^a}\frac{\partial X'^b}{\partial X'^c}\partial_d A^c. \tag{5.12}$$

Likewise we can work out $g'_{ab,c}$

$$g'_{ab,c} = \frac{\partial X^k}{\partial X'^a}\frac{\partial X^l}{\partial X'^b}\frac{\partial X^m}{\partial X'^c}g_{kl,m} + g_{kl}\frac{\partial X^l}{\partial X'^b}\frac{\partial\partial X^k}{\partial X'^a\partial X'^c}$$
$$+ g_{kl}\frac{\partial X^k}{\partial X'^a}\frac{\partial\partial X^l}{\partial X'^b\partial X'^c}. \tag{5.13}$$

The first term contains the homogeneous transformation part, the remaining terms are the inhomogeneous parts. Taking the definition of the Christoffel symbol into account, there are three terms of the form $g_{ab,c}$. So there will be a total of six inhomogeneous terms in the transformed Christoffel symbol, four of which have plus signs and two of which have negative signs. Four of these terms cancel each other and the remaining two inhomogeneous terms are added up and cancel the factor $1/2$ in the definition of the Christoffel symbol. This yields the result.

Exercise 1.9. We have $A'^j = \partial X'^j/\partial X^s A^s$ which we can simply apply to both sides of Eq. (1.238). Then $(\partial X^c/\partial X'^j)(\partial X'^j/\partial X^s) A^s = \delta^c_s A^s = A^c$ and likewise for the second term which implies the result.

Exercise 1.10. We showed that $ds^2 = dx^2 + dy^2 = dr^2 + r^2 d\varphi^2$ in polar coordinates. Then a direct substitution gives

$$ds^2 = 4\frac{dx^2 + dy^2}{(1 - x^2 - y^2)^2} = 4\frac{dr^2 + r^2 d\varphi^2}{(1 - r^2)^2}. \tag{5.14}$$

Exercise 1.11. Start with $\rho = 2r/(1 - r^2)$ which is a quadratic equation in r given by $r^2 + 2r/\rho - 1 = 0$ which we can solve to find $r = (\sqrt{1 + \rho^2} - 1)/\rho$ (the other solution is unphysical). Next we compute $d\rho$ for which we have

$$d\rho = 2\frac{1 + r^2}{(1 - r^2)^2} dr \quad d\rho^2 = 4\frac{(1 + r^2)^2}{(1 - r^2)^4} dr^2, \tag{5.15}$$

so that we can write

$$4\frac{dr^2}{(1 - r^2)^2} dr^2 = \frac{(1 - r^2)^2}{(1 + r^2)^2} d\rho^2. \tag{5.16}$$

Once we rewrite the right-hand side in terms of ρ we are done.

$$1 - r^2 = 1 - \frac{(\sqrt{1 + \rho^2} - 1)^2}{\rho^2} = 2\frac{\sqrt{1 + \rho^2} - 1}{\rho^2}, \tag{5.17}$$

$$1 + r^2 = 2\frac{1 + \rho^2 - \sqrt{1 + \rho^2}}{\rho^2}. \tag{5.18}$$

The final calculation is

$$\frac{(1 - r^2)^2}{(1 + r^2)^2} = \frac{(\sqrt{1 + \rho^2} - 1)^2}{(1 + \rho^2 - \sqrt{1 + \rho^2})^2}$$

$$= \frac{1 + \rho^2 + 1 - 2\sqrt{1 + \rho^2}}{(1 + \rho^2)^2 + (1 + \rho^2) - 2(1 + \rho^2)\sqrt{1 + \rho^2}}$$

$$= \frac{1 + \rho^2 + 1 - 2\sqrt{1 + \rho^2}}{(1 + \rho^2)\left[(1 + \rho^2) + 1 - 2\sqrt{1 + \rho^2}\right]}$$

$$= \frac{1}{1 + \rho^2}. \tag{5.19}$$

Hence we find the desired result

$$ds^2 = 4\frac{dr^2 + r^2 d\varphi^2}{(1 - r^2)^2} = \frac{d\rho^2}{1 + \rho^2} + \rho^2 d\varphi^2. \tag{5.20}$$

With $\rho = \sinh\chi$ we find $d\rho = \cosh\chi\, d\chi$ and $d\rho^2 = \cosh^2\chi\, d\chi^2$. Next $1 + \rho^2 = 1 + \sinh^2\chi = \cosh^2\chi$ where we used the standard hyperbolic identity. Therefore, $ds^2 = d\chi^2 + \sinh^2\chi d\varphi^2$.

Exercise 1.12. We start with $\nabla_i A^i = \partial_i A^i + \Gamma^i_{ij} A^j$ and now recall Eq. (1.100) so that we can write

$$\nabla_i A^i = \partial_i A^i + \frac{1}{\sqrt{-g}}\partial_i \sqrt{-g} A^i$$

$$= \frac{1}{\sqrt{-g}}\left(\sqrt{-g}\partial_i A^i + A^i \partial_i \sqrt{-g}\right)$$

$$= \frac{1}{\sqrt{-g}}\partial_i \left(\sqrt{-g} A^i\right). \tag{5.21}$$

Exercise 1.13. The metric tensor and inverse metric tensor are given by

$$g_{ij} = \begin{pmatrix} 1 & 0 \\ 0 & \cosh^{-2}z \end{pmatrix}, \quad g^{ij} = \begin{pmatrix} 1 & 0 \\ 0 & \cosh^2 z \end{pmatrix}. \tag{5.22}$$

Hence, $\partial_y g_{ij} = 0$. The only non-vanishing Christoffel symbol component is $\Gamma^2_{22} = -2\tanh z$. The Lagrangian is $L = \dot{y}^2 + \dot{z}^2/\cosh^4 z$ so that $\ddot{y} = 0$ and $y = c\lambda + d$ (c, d are constants), while the z equation is

$$-4\dot{z}^2\frac{\sinh z}{\cosh^5 z} = \frac{d}{d\lambda}\frac{2\dot{z}}{\cosh^4 z} = \frac{2\ddot{z}}{\cosh^4 z} - 8\dot{z}^2\frac{\sinh z}{\cosh^5 z}. \tag{5.23}$$

Therefore, we arrive at

$$\ddot{z} - 2\dot{z}^2\frac{\sinh z}{\cosh z} = 0, \tag{5.24}$$

and can verify that the only non-vanishing Christoffel symbol component is Γ^2_{22}. Now, we divide this last equation by $\dot{z}$ and write

$$\frac{\ddot{z}}{\dot{z}} = 2\dot{z}\frac{\sinh z}{\cosh z}. \tag{5.25}$$

We note that

$$\frac{d}{d\lambda}\log(\dot{z}) = \frac{\ddot{z}}{\dot{z}}, \tag{5.26}$$

$$2\frac{d}{d\lambda}\log(\cosh z) = 2\dot{z}\frac{\sinh z}{\cosh z}, \tag{5.27}$$

and can now integrate both sides with respect λ which gives

$$\log(\dot{z}) = \log(\cosh^2 z) + C, \quad \Rightarrow \quad \dot{z} = \tilde{C}\cosh^2 z. \tag{5.28}$$

Separation of variables leads to

$$\frac{dz}{\cosh^2 z} = \tilde{C}d\lambda, \quad \Rightarrow \quad \tanh z = \tilde{C}\lambda + \tilde{D}. \tag{5.29}$$

Therefore, using $y = c\lambda + d$ we can eliminate λ and arrive at

$$\tanh z = \frac{\tilde{C}}{c}(y - d) + \tilde{D}, \tag{5.30}$$

which is the equation of straight line. This also strongly suggests that one should introduce a new coordinate of the form $x = \tanh z$. One can check

$$dx = \frac{1}{\cosh^2 z}dz \quad dx^2 = \frac{1}{\cosh^4 z}dz^2, \tag{5.31}$$

so that

$$ds^2 = dx^2 + dy^2 = \frac{dz^2}{\cosh^4 z} + dy^2. \tag{5.32}$$

Exercise 1.14. Write out (1.114) + (1.115) − (1.116) in detail

$$\partial_a g_{bc} + \partial_b g_{ca} - \partial_c g_{ab} - C^d_{ab}g_{dc} - C^d_{ba}g_{cd}$$
$$+ C^d_{ca}g_{db} - C^d_{ac}g_{bd} + C^d_{cb}g_{ad} - C^d_{bc}g_{da} = 0. \tag{5.33}$$

Now $C^d_{ca} - C^d_{ac} = 2T_{ca}{}^d$, $C^d_{cb} - C^d_{bc} = 2T_{cb}{}^d$ and $-C^d_{ab} - C^d_{ba} = -2C^d_{ab} + 2T_{ab}{}^d$. So we get

$$C^d_{ab}g_{dc} = \frac{1}{2}\left(\partial_a g_{bc} + \partial_b g_{ca} - \partial_c g_{ab}\right) + T_{ab}{}^d g_{dc} + T_{ca}{}^d g_{bd} + T_{cb}{}^d g_{da}. \tag{5.34}$$

Finally we apply g^{ic} to both sides and arrive at

$$C^i_{ab} = \Gamma^i_{ab} + T_{ab}{}^i + T^i{}_{ab} + T^i{}_{ba} = \Gamma^i_{ab} + T_{ab}{}^i - T_b{}^i{}_a + T^i{}_{ab}, \tag{5.35}$$

which is the stated result.

Exercise 1.15. The metric and inverse metric always satisfy $g_{bc}g^{cd} = \delta^d_b$. Let us differentiate this using $\tilde{\nabla}_a$. This gives

$$\tilde{\nabla}_a\left(g_{bc}g^{cd}\right) = 0,$$

$$\left(\tilde{\nabla}_a g_{bc}\right)g^{cd} + g_{bc}\left(\tilde{\nabla}_a g^{cd}\right) = 0, \tag{5.36}$$

$$Q_a g_{bc}g^{cd} + g_{bc}\left(\tilde{\nabla}_a g^{cd}\right) = 0,$$

where in the second step the product rule was used. Now, we apply g^{bf} and arrive at

$$g^{bf}g_{bc}\left(\tilde{\nabla}_a g^{cd}\right) = -Q_a g_{bc}g^{cd}g^{bf},$$

$$\delta^f_c\left(\tilde{\nabla}_a g^{cd}\right) = -Q_a \delta^f_c g^{cd},$$

$$\tilde{\nabla}_a g^{fd} = -Q_a g^{fd}. \tag{5.37}$$

Curvature

Exercise 1.16. Following Example 1.14, our coordinates are $X^i = (\theta, \phi)$ and the non-vanishing Christoffel symbol components are $\Gamma^1_{22} = -\sin\theta\cos\theta$, $\Gamma^2_{12} = \cot\theta$. Next, we can recall the equations for parallel transport (1.128). T^a is the tangent vector of the curve along which we transport.

(i) Two points with the same longitude have coordinates (θ_1, ϕ_0) and (θ_2, ϕ_0), respectively. A curve C connecting these points is given by $X^a(\lambda) = (\theta_1 + \lambda(\theta_2 - \theta_1), \phi_0)$ and has tangent vector $T^a = (\theta_2 - \theta_1, 0)$ which we write as $T^a = (\bar{\theta}, 0)$. Therefore, the equations for parallel transport are

$$\frac{dV^1}{d\lambda} = -\Gamma^1_{ac}T^aV^c = -\Gamma^1_{22}T^2V^2 = 0, \tag{5.38}$$

$$\frac{dV^2}{d\lambda} = -\Gamma^2_{12}(T^1V^2 + T^2V^1) = -\bar{\theta}\cot\theta V^2. \tag{5.39}$$

The first equation simply yields $V^1 = V^1_{\text{initial}}$, in the second equation we can separate variables. We use $\theta(\lambda) = \theta_1 + \bar{\theta}\lambda$ so that

$$\frac{dV^2}{V^2} = -\bar{\theta}\cot\left(\theta_1 + \bar{\theta}\lambda\right)d\lambda, \tag{5.40}$$

which we can integrate to find

$$\log(V^2) = -\log\left(\sin\left(\theta_1 + \bar{\theta}\lambda\right)\right) + C_1, \tag{5.41}$$

$$\Rightarrow \quad V^2 = \frac{\tilde{C}_1}{\sin\left(\theta_1 + \bar{\theta}\lambda\right)}. \tag{5.42}$$

We can determine the value of the constant of integration by the condition $V^2(0) = V^2_{\text{initial}}$ and arrive at the result

$$V^1 = V^1_{\text{initial}}, \tag{5.43}$$

$$V^2 = \frac{\sin(\theta_1)}{\sin\left(\theta_1 + \bar{\theta}\lambda\right)}V^2_{\text{initial}}. \tag{5.44}$$

(ii) We now repeat this process for parallel transport along constant latitudes. A curve C connecting two such points is given by $X^a(\lambda) = (\theta_0, \phi_1 + \lambda(\phi_2 - \phi_1))$ and has tangent vector $T^a = (0, \phi_2 - \phi_1)$ which we write as $T^a = (0, \bar{\phi})$. The parallel transport equations are

$$\frac{dV^1}{d\lambda} = -\Gamma^1_{ac}T^aV^c = -\Gamma^1_{22}T^2V^2 = \bar{\phi}\sin\theta_0\cos\theta_0 V^2, \tag{5.45}$$

$$\frac{dV^2}{d\lambda} = -\Gamma^2_{12}(T^1 V^2 + T^2 V^1) = -\bar{\phi}\cot\theta_0 V^1. \tag{5.46}$$

This pair of equations is the harmonic oscillator equation in first-order formulation. To see this, differentiate the second equation with respect to λ and eliminate V^1. This gives

$$\frac{d^2 V^2}{d\lambda^2} = -\bar{\phi}\cot\theta_0 \frac{dV^1}{d\lambda} = -\bar{\phi}^2 \cos^2\theta_0 V^2 = -k^2 V^2, \tag{5.47}$$

with $k = \bar{\phi}\cos\theta_0$. The solution to this is

$$V^2 = C_1 \sin(k\lambda) + C_2 \cos(k\lambda), \tag{5.48}$$

which implies that V^1 is given by

$$V^1 = -\frac{\tan\theta_0}{\bar{\phi}}\frac{dV^2}{d\lambda} = \sin\theta_0(C_2 \sin(k\lambda) - C_1 \cos(k\lambda)). \tag{5.49}$$

We can now fix our constants of integration by $V^i(0) = V^i_{\text{initial}}$. This gives the second result

$$V^1 = V^1_{\text{initial}} \cos(k\lambda) + V^2_{\text{initial}} \sin\theta_0 \sin(k\lambda), \tag{5.50}$$

$$V^2 = V^2_{\text{initial}} \cos(k\lambda) - \frac{V^1_{\text{initial}}}{\sin\theta_0}\sin(k\lambda). \tag{5.51}$$

(iii) Let us begin by transporting $V_{\text{initial}} = (1,0)$ from $P_1 = (\pi/2, 0)$ to $P_2 = (\epsilon, 0)$, ϕ is constant and we find

$$V^1(P_2) = V^1_{\text{initial}} = 1, \tag{5.52}$$

$$V^2(P_2) = \frac{\sin(\pi/2)}{\sin(\epsilon)}V^2_{\text{initial}} = 0. \tag{5.53}$$

We cannot transport to the north pole as our coordinates become singular at this point. Next, we transport from $P_2 = (\epsilon, 0)$ to $P_3 = (\epsilon, \pi/2)$, now θ is constant, $k = \bar{\phi}\cos\theta_0 = \pi/2\cos(\epsilon)$ hence

$$V^1(P_3) = V^1(P_2)\cos(\pi/2\cos(\epsilon)) = \cos(\pi/2\cos(\epsilon)), \tag{5.54}$$

$$V^2(P_3) = \frac{V^1(P_2)}{\sin(\epsilon)} \sin(\pi/2 \cos(\epsilon)) = \frac{\sin(\pi/2 \cos(\epsilon))}{\sin(\epsilon)}. \quad (5.55)$$

Next, we transport from $P_3 = (\epsilon, \pi/2)$ to $P_4 = (\pi/2, \pi/2)$. This leads to

$$V^1(P_4) = V^1(P_3) = \cos(\pi/2 \cos(\epsilon)), \quad (5.56)$$

$$V^2(P_4) = \frac{\sin(\epsilon)}{\sin(\pi/2)} V^2(P_3) = \frac{\sin(\pi/2 \cos(\epsilon))}{\sin(\pi/2)}. \quad (5.57)$$

We can now safely consider the limit $\epsilon \to 0$ which yields

$$V^1(P_4) = 0, \quad (5.58)$$

$$V^2(P_4) = 1. \quad (5.59)$$

Finally, we transport this vector from $P_4 = (\pi/2, \pi/2)$ back to $P_1 = (\pi/2, 0)$, this gives $k = \bar{\phi} \cos \theta_0 = 0$ and we arrive at the result

$$V^1_{\text{final}} = V^2(P_4) \sin(k) = 0, \quad (5.60)$$

$$V^2_{\text{final}} = V^2(P_4) \cos(k) = 1. \quad (5.61)$$

This shows that the final vector is rotated by $90°$.

Exercise 1.17. We begin with the Christoffel symbol of the metric defined by $ds^2 = v^2 du^2 + u^2 dv^2$ which is given by

$$\Gamma^1_{11} = 0, \quad \Gamma^1_{12} = \frac{1}{v}, \quad \Gamma^1_{22} = -\frac{u}{v^2},$$

$$\Gamma^2_{11} = -\frac{v}{u^2}, \quad \Gamma^2_{12} = \frac{1}{u}, \quad \Gamma^2_{22} = 0. \quad (5.62)$$

We compute R_{1212} by direct calculation using Eq. (1.164) which in this case is

$$R_{121}{}^2 = \partial_2 \Gamma^2_{11} - \partial_1 \Gamma^2_{21} + \Gamma^2_{2c} \Gamma^c_{11} - \Gamma^2_{1c} \Gamma^c_{21}. \quad (5.63)$$

Therefore,

$$\partial_2 \Gamma^2_{11} = -\frac{1}{u^2}, \quad \partial_1 \Gamma^2_{21} = -\frac{1}{u^2},$$

$$\Gamma^2_{2c}\Gamma^c_{11} = \Gamma^2_{21}\Gamma^1_{11} + \Gamma^2_{22}\Gamma^2_{11} = 0,$$
(5.64)

$$\Gamma^2_{1c}\Gamma^c_{21} = \Gamma^2_{11}\Gamma^1_{21} + \Gamma^2_{12}\Gamma^2_{21} = -\frac{v}{u^2}\frac{1}{v} + \frac{1}{u}\frac{1}{u} = 0,$$
(5.65)

and we find the desired result $R_{1212} = 0$.

Exercise 1.18. We begin with

$$ds^2 = v^2 du^2 + u^2 dv^2 = u^2 v^2 \left[\frac{du^2}{u^2} + \frac{dv^2}{v^2}\right] = e^{2U} e^{2V} \left[dU^2 + dV^2\right],$$
(5.66)

where we used $u = \exp U$ and $v = \exp V$. Next we introduce the new coordinates $V + U = \alpha$, $V - U = \beta$ so that $V = (\alpha + \beta)/2$ and $U = (\alpha - \beta)/2$. Substitution into the line element gives

$$ds^2 = e^{2U} e^{2V} \left[dU^2 + dV^2\right] = e^{2\alpha}\frac{1}{2}\left[d\alpha^2 + d\beta^2\right].$$
(5.67)

Next, we introduce

$$r = \frac{1}{\sqrt{2}}e^{\alpha}, \quad dr = \frac{1}{\sqrt{2}}e^{\alpha}d\alpha,$$
(5.68)

and we arrive at

$$ds^2 = e^{2\alpha}\frac{1}{2}\left[d\alpha^2 + d\beta^2\right] = dr^2 + r^2 d\beta^2,$$
(5.69)

which is Euclidean space in polar coordinates, see Example 1.2. Putting all the transformations together, you hopefully arrive at

$$x = \frac{1}{\sqrt{2}}uv\cos\left(\log\frac{u}{v}\right), \quad y = \frac{1}{\sqrt{2}}uv\sin\left(\log\frac{u}{v}\right).$$
(5.70)

Finally, it is also a good exercise to show that $dx^2 + dy^2$ indeed produces the required metric.

Exercise 1.19. We write out property (iii) four times as follows:

$$W_{\underline{abcd}} + W_{cabd} + W_{bcad} = 0, \tag{5.71}$$

$$W_{dabc} + W_{bdac} + W_{\underline{abdc}} = 0, \tag{5.72}$$

$$W_{\underline{cdab}} + W_{acdb} + W_{dacb} = 0, \tag{5.73}$$

$$W_{bcda} + W_{dbca} + W_{\underline{cdba}} = 0, \tag{5.74}$$

where we underlined all terms which are part of the identity we wish to prove. Now we compute (5.71) − (5.72) − (5.73) + (5.74). This yields

$$2W_{\underline{abcd}} - 2W_{\underline{cdab}} + W_{cabd} + W_{bcad} - W_{dabc} - W_{bdac} \tag{5.75}$$

$$-W_{acdb} - W_{dacb} + W_{bcda} + W_{dbca} = 0, \tag{5.76}$$

where property (ii) was used. All the remaining terms will now cancel. For instance

$$W_{cabd} - W_{acdb} = -W_{acbd} - W_{acdb} = W_{acdb} - W_{acdb} = 0,$$

$$W_{bcad} + W_{bcda} = W_{bcad} - W_{bcad} = 0, \tag{5.77}$$

and so on for the remaining two pairs. Hence $2W_{abcd} - 2W_{cdab} = 0$, which we wanted to show.

Exercise 1.20. One possible proof is as follows: The Riemann curvature tensor is skew-symmetric in the first and last pair of indices. The first pair can take $(n-1)n/2$ different values, and so can the second pair, so we write R_{AB} with $A, B = \{1, \ldots, (n-1)n/2\}$ for R_{abcd}. Next, we know that $R_{abcd} = R_{cdab}$ which becomes $R_{AB} = R_{BA}$, in other words R_{AB} is symmetric and therefore has $((n-1)n/2)((n-1)n/2 + 1)/2$ independent components. We are left with property (iii) to take into account, this is not easy, and involves some combinatorics. There are 'n choose 4' equations we need to take into account. This gives

$$
\left(\frac{(n-1)n}{2}\right)\left(\frac{(n-1)n}{2}+1\right) - \binom{n}{4}
$$

$$
= \frac{1}{4}(n-1)n\left(\frac{(n-1)n+2}{2}\right) - \frac{n!}{(n-4)!\,4!}
$$

$$
= \frac{1}{8}(n-1)n[(n-1)n+2] - \frac{n(n-1)(n-2)(n-3)}{24}, \qquad (5.78)
$$

which after a bit of simplification gives the result $n^2(n^2-1)/12$.

Note that the derivation involves a Binomial coefficient which is not defined for $n < 4$. In the cases $n = 2$ and $n = 3$, property (iii) simply does not give any additional equations.

5.2.　Solutions: Einstein Field Equations

Some physics background

Exercise 2.1. We begin with the Euler equation for hydrostatic equilibrium which is $\rho g = \nabla p$. Next, assuming spherical symmetry we have $dp/dr = -\rho Gm/r^2$. The mass m and the density ρ are related by the standard relation $dm/dr = 4\pi\rho r^2$. Since we consider a constant density star $\rho = \rho_0 = \text{const.}$, we have $m = (4\pi/3)\rho_0 r^3$. Therefore, $dp/dr = -(4\pi G/3)\rho_0^2 r$, which we can integrate to find

$$
p(r) = p_c - \frac{4\pi G}{6}\rho_0^2 r^2, \qquad (5.79)
$$

where p_c is a constant of integration which corresponds to the central pressure of the Newtonian star. One verifies that the units of $G\rho_0^2 r^2$ are those of pressure, which has units of force per area.

Next, we need to find the boundary of the star which we choose to be the vanishing pressure surface. Hence, we define R such that

$p(R) = 0$. This gives this equation

$$p_c = \frac{4\pi G}{6} \rho_0^2 R^2, \tag{5.80}$$

while the compactness measure is $M/R = (4\pi/3)\rho_0 R^2$. For a Newtonian star there are no constraints limiting the mass and size of this hypothetical star. The general relativistic treatments of static and spherically symmetric stars (Sec. 3.3) with constant density will show that $M/R < 4/9$ which makes these solutions more realistic.

Exercise 2.2. Let us begin with $i = 1, j = 2, k = 3$, then $\partial_1 F_{23} + \partial_3 F_{12} + \partial_2 F_{31} = 0$. Using the components of F_{ij} means that this is $\partial_1(-B_x) + \partial_3(-B_z) + \partial_2(-B_y) = 0$ which is the explicit form of div $\boldsymbol{B} = 0$. For all other index combinations where i, j, k take values $1, 2, 3$ we either arrive at a trivial identity or the same equation. This is the second equation of (2.23).

Now, let us choose $i = 0, j = 1, k = 2$, then $\partial_0 F_{12} + \partial_2 F_{01} + \partial_1 F_{20} = 0$. Again, using the components we have $\partial_0(-B_z) + \partial_2 E_x + \partial_1(-E_y) = 0$ or

$$\partial_1 E_y - \partial_2 E_x + \frac{1}{c}\frac{\partial B_z}{\partial t}. \tag{5.81}$$

We recognise the first term as the z-component of the vector curl $\boldsymbol{E}$, therefore we have one of the three equations. The choices $i = 0$, $j = 1, k = 3$ and $i = 0, j = 2, k = 3$ yield the y and x components, respectively. Hence, the first equation of (2.23) is shown and therefore the equivalence established.

Exercise 2.3. Let us start with $\eta_{ij}u^i u^j = -1$. In the new coordinates, we should write $\eta_{ij}u^i u^j = -c^2$ and find $u^i = (c, 0, 0, 0)$, similar to before but multiplied by c. The physical meaning of the components of the energy–momentum tensor is independent of the choice of coordinates, nonetheless, we need to compensate for factors of c from the Minkowski metric. We can write $T^{ij} = (\rho + p/c^2)u^i u^j + p\eta^{ij}$, and can check explicitly, $T^{00} = (\rho + p/c^2)c^2 + p(-1) = \rho c^2$, and $T^{11} = T^{22} = T^{33} = p$. In general, we would write $u^i = (c, \boldsymbol{v})$ which is nice in the sense that this vector has units of velocity. Both formulations are related by the simple transformation $u^i \to cu^i$. We should also note that the choice of coordinates $X^i = (t, x, y, z)$ means that time and space have the same units. This gives a first hint that setting $c = 1$ is a pretty good idea as it is very easy to get factors of c wrong.

Exercise 2.4. We can read off the matrix form directly from the coordinate transformation

$$L^a{}_b = \begin{pmatrix} \cosh(\zeta) & -\sinh(\zeta) & 0 & 0 \\ -\sinh(\zeta) & \cosh(\zeta) & 0 & 0 \\ 0 & 0 & 1 & 0 \\ 0 & 0 & 0 & 1 \end{pmatrix}. \tag{5.82}$$

To compute the determinant, we expand along the fourth and then along the third row which gives

$$\det(L^a{}_b) = \det \begin{pmatrix} \cosh(\zeta) & -\sinh(\zeta) \\ -\sinh(\zeta) & \cosh(\zeta) \end{pmatrix} = \cosh^2(\zeta) - \sinh^2(\zeta) = 1. \tag{5.83}$$

Next, we need to find $L^c{}_a L^d{}_b \eta_{cd}$. First, we work out

$$L^d{}_b \eta_{cd} = \begin{pmatrix} -\cosh(\zeta) & -\sinh(\zeta) & 0 & 0 \\ \sinh(\zeta) & \cosh(\zeta) & 0 & 0 \\ 0 & 0 & 1 & 0 \\ 0 & 0 & 0 & 1 \end{pmatrix}, \tag{5.84}$$

and next one can verify $L^c{}_a(L^d{}_b\eta_{cd}) = \eta_{ab}$ which is what we wanted to show.

Comparing Eq. (5.82) with Eq. (2.13) gives the following identifications $\cosh(\zeta) = \gamma$ and $\sinh(\zeta) = \gamma\beta$, and then $\sinh(\zeta)/\cosh(\zeta) = \tanh(\zeta) = \beta = v/c$. The angle α in Fig. 2.1 is related to the velocity β by $\beta = \tan(\alpha)$ which gives $\tan(\alpha) = \tanh(\zeta)$ and hence $\alpha = \arctan(\tanh(\zeta))$.

Exercise 2.5. A 2D rotation matrix $R(\theta)$ is generally written as

$$R(\theta) = \begin{pmatrix} \cos(\theta) & -\sin(\theta) \\ \sin(\theta) & \cos(\theta) \end{pmatrix}, \tag{5.85}$$

so that $R(i\theta)$ is given by

$$R(i\theta) = \begin{pmatrix} \cos(i\theta) & -\sin(i\theta) \\ \sin(i\theta) & \cos(i\theta) \end{pmatrix} = \begin{pmatrix} \cosh(\theta) & -i\sinh(\theta) \\ i\sinh(\theta) & \cosh(\theta) \end{pmatrix}. \tag{5.86}$$

Now, we can compute the transformation property

$$R(i\theta)\begin{pmatrix} ict \\ x \end{pmatrix} = \begin{pmatrix} \cosh(\theta) & -i\sinh(\theta) \\ i\sinh(\theta) & \cosh(\theta) \end{pmatrix}\begin{pmatrix} ict \\ x \end{pmatrix} \tag{5.87}$$

$$= \begin{pmatrix} \cosh(\theta)ict - i\sinh(\theta)x \\ i\sinh(\theta)ict + \cosh(\theta)x \end{pmatrix} \tag{5.88}$$

$$= \begin{pmatrix} i(\cosh(\theta)ct - \sinh(\theta)x) \\ -\sinh(\theta)ct + \cosh(\theta)x \end{pmatrix} = \begin{pmatrix} ict' \\ x' \end{pmatrix}, \tag{5.89}$$

which is in agreement with the hyperbolic form of the transformations given by Eqs. (2.133) and (2.134). This justifies the notion of hyperbolic rotations for Lorentz boosts.

Exercise 2.6. We begin with $L^a{}_b$ from Eq. (5.82) with ζ_1 and ζ_2, respectively, and compute

$$L^a{}_b(\zeta_1)L^b{}_c(\zeta_2)$$

$$= \begin{pmatrix} -\cosh(\zeta_1) & -\sinh(\zeta_1) & 0 & 0 \\ \sinh(\zeta_1) & \cosh(\zeta_1) & 0 & 0 \\ 0 & 0 & 1 & 0 \\ 0 & 0 & 0 & 1 \end{pmatrix}$$

$$\times \begin{pmatrix} -\cosh(\zeta_1) & -\sinh(\zeta_1) & 0 & 0 \\ \sinh(\zeta_1) & \cosh(\zeta_1) & 0 & 0 \\ 0 & 0 & 1 & 0 \\ 0 & 0 & 0 & 1 \end{pmatrix}$$

$$= \begin{pmatrix} -\cosh(\zeta_1 + \zeta_2) & -\sinh(\zeta_1 + \zeta_2) & 0 & 0 \\ \sinh(\zeta_1 + \zeta_2) & \cosh(\zeta_1 + \zeta_2) & 0 & 0 \\ 0 & 0 & 1 & 0 \\ 0 & 0 & 0 & 1 \end{pmatrix}, \quad (5.90)$$

where we used the hyperbolic identities

$$\cosh(\zeta_1)\cosh(\zeta_2) + \sinh(\zeta_1)\sinh(\zeta_2) = \cosh(\zeta_1 + \zeta_2), \quad (5.91)$$

$$\cosh(\zeta_1)\sinh(\zeta_2) + \sinh(\zeta_1)\cosh(\zeta_2) = \sinh(\zeta_1 + \zeta_2). \quad (5.92)$$

Therefore, we see that the rapidity of the overall boost is $\zeta = \zeta_1 + \zeta_2$.

We can use this result and combine it with the previous exercise where we showed that $\tanh(\zeta_1) = \beta_1 = v_1/c$. Since $\zeta = \zeta_1 + \zeta_2$ we also have that $\tanh(\zeta) = \tanh(\zeta_1 + \zeta_2)$. Now apply the hyperbolic identity which expands $\tanh(\zeta_1 + \zeta_2)$, this yields

$$\tanh(\zeta) = \tanh(\zeta_1 + \zeta_2) = \frac{\tanh(\zeta_1) + \tanh(\zeta_2)}{1 + \tanh(\zeta_1)\tanh(\zeta_2)}. \quad (5.93)$$

Finally, we can replace all hyperbolic tangent terms with their respective velocities and arrive at the desired result

$$\beta = \frac{\beta_1 + \beta_2}{1 + \beta_1\beta_2}, \quad \Leftrightarrow \quad v/c = \frac{v_1/c + v_2/c}{1 + (v_1/c)(v_2/c)}$$

$$\Rightarrow \quad v = \frac{v_1 + v_2}{1 + v_1 v_2/c^2}. \tag{5.94}$$

Exercise 2.7. We already computed the T^{00} component. Next, let us look at the $T^{0\alpha}$ components where $\alpha = 1, 2, 3$ can only take the spatial values. Hence,

$$T^{0\alpha} = -\frac{1}{4\pi}\left(F^{i0}F_i{}^\alpha - \frac{1}{4}\eta^{0\alpha}F^{mn}F_{mn} \right)$$

$$= -\frac{1}{4\pi}F^{i0}F_i{}^\alpha = -\frac{1}{4\pi}F^{\beta 0}F_\beta{}^\alpha, \tag{5.95}$$

because $\eta^{0\alpha} = 0$ for all α, and $F^{00} = 0$.

Therefore, we find

$$T^{01} = -\frac{1}{4\pi}F^{\beta 0}F_\beta{}^1 = -\frac{1}{4\pi}(F^{20}F_2{}^1 + F^{30}F_3{}^1)$$

$$= -\frac{1}{4\pi}(E_y B_z + E_z(-B_y)) = \frac{1}{4\pi}(E_z B_y - E_y B_z), \tag{5.96}$$

which we identify as the x-component of the vector product of $\boldsymbol{E}$ and $\boldsymbol{B}$ and we suspect that $cT^{0\alpha} = S^\alpha$. A direct calculation will show that this is indeed the case for the other two components as well.

The spatial components of the energy–momentum tensor are given by

$$T_{\alpha\beta} = -\frac{1}{4\pi}\left(E_\alpha E_\beta + B_\alpha B_\beta - \frac{1}{2}\eta_{\alpha\beta}(|\boldsymbol{E}|^2 + |\boldsymbol{B}|^2) \right). \tag{5.97}$$

Exercise 2.8. We work in Minkowski space so $g_{ij} = \eta_{ij}$, and recall $F_{ab} = \partial_a A_b - \partial_b A_a$. We will make the metric in the action explicit and write $\eta^{ac}\eta^{bd}F_{ab}F_{cd}$ so that

$$S = \int \frac{1}{\alpha}\eta^{ac}\eta^{bd}F_{ab}F_{cd} + A_b J^b \, d^4x. \qquad (5.98)$$

Considering $A_a + \delta A_a$ gives

$$\delta S = \int \frac{2}{\alpha}(\partial_a \delta A_b - \partial_b \delta A_a)F^{ab} + \delta A_b J^b \, d^4x$$

$$= \int -\frac{2}{\alpha}\partial_a F^{ab}\delta A_b + \frac{2}{\alpha}\partial_b F^{ab}\delta A_a + \delta A_b J^b \, d^4x$$

$$= \int \left[\frac{4}{\alpha}\partial_b F^{ab} + J^a\right]\delta A_a \, d^4x, \qquad (5.99)$$

where we integrated by parts and neglected the boundary terms.

Therefore, we arrive at the following equations $\partial_b F^{ab} = -(\alpha/4)J^a$ which matches the Maxwell equations (2.27) provided we choose $\alpha = 16\pi/c$.

Geometry and gravity

Exercise 2.9. The easiest way to see this is to use Eq. (2.76) as the Einstein field equation. Then $T_{ij} = 0$ implies $T = 0$ and therefore the vacuum field equations are $R_{ij} = 0$. Taking the trace gives $R = 0$ as desired.

Exercise 2.10. As mentioned in the text, we will show this directly by computing the covariant derivative of the left-hand side. This is

$$\nabla^i(G_{ij} + \Lambda g_{ij}) = \nabla^i G_{ij} + \Lambda \nabla^i g_{ij}, \qquad (5.100)$$

where we used that Λ is a constant and thus not affected by the derivative. Next, by Theorem 1.6 we have $\nabla^i G_{ij} = 0$. Moreover, we assume the covariant derivative to be metric compatible, therefore $\nabla^i g_{ij} = 0$ and consequently the field equations with cosmological term also imply the energy–momentum conservation equation.

Exercise 2.11. We begin with applying g^{ij} to Eq. (2.77). This gives $R - 2R + 4\Lambda = \kappa T$, hence $-R = \kappa T - 4\Lambda$ which we substitute back into Eq. (2.77). This gives

$$R_{ij} + \frac{1}{2}g_{ij}\left(\kappa T - 4\Lambda\right) + \Lambda g_{ij} = \kappa T_{ij},$$

$$R_{ij} - \Lambda g_{ij} = \kappa T_{ij} - \kappa\frac{1}{2}g_{ij}T = \kappa\left(T_{ij} - \frac{1}{2}g_{ij}T\right),$$

$$\Rightarrow \quad R_{ij} - \Lambda g_{ij} = \kappa\left(T_{ij} - \frac{1}{2}Tg_{ij}\right). \quad (5.101)$$

Exercise 2.12. We begin by writing out explicitly the covariant derivative

$$\nabla_i\mathcal{F}^{ij} = \partial_i\mathcal{F}^{ij} + \Gamma^i_{ik}\mathcal{F}^{kj} + \Gamma^j_{ik}\mathcal{F}^{ik}$$

$$= \partial_i\mathcal{F}^{ij} + \Gamma^i_{ik}\mathcal{F}^{kj}, \quad (5.102)$$

where we used $\Gamma^j_{ik}\mathcal{F}^{ik} = 0$ because Γ is symmetric in the two lower indices while $\mathcal{F}$ is skew-symmetric. Using Eq. (1.100), we have

$$\nabla_i\mathcal{F}^{ij} = \partial_i\mathcal{F}^{ij} + \frac{1}{\sqrt{|g|}}\partial_k\left(\sqrt{|g|}\right)\mathcal{F}^{kj}. \quad (5.103)$$

Next, we can multiply this equation by $\sqrt{|g|}$ and apply the product rule. This shows that $\nabla_i\mathcal{F}^{ij} = 0$ is equivalent to

$$\partial_k\left(\sqrt{|g|}\mathcal{F}^{kj}\right) = 0. \quad (5.104)$$

Integration over the entire space results in a conserved quantity.

Exercise 2.13. One could suggest that the simplest metric component is

$$1 \pm \frac{2GM}{c^2 r} + \alpha\Lambda r^2. \quad (5.105)$$

The sign here does not matter as both α and Λ are not assumed to be positive.

The gravitational field g is related to the Christoffel symbol, hence, the simplest approach towards a Λ-corrected Newtonian force law would be to differentiate our suggested metric component with respect to r which gives

$$-\frac{2GM}{c^2 r^2} + 2\alpha\Lambda r. \tag{5.106}$$

Recall that the definition of the Christoffel symbol contains a factor of $1/2$, and also the minus sign when relating the potential to the field. Taking these into account leads to

$$\frac{1}{c^2}g = \frac{GM}{c^2 r^2} - \alpha\Lambda r \quad \text{or} \quad g = \frac{GM}{r^2} - \alpha c^2 \Lambda r. \tag{5.107}$$

Therefore, the cosmological constant contributes linearly to the Newtonian force law.

Within the Solar System Newtonian gravity is a very good approximation to the gravitational field of the Sun. The dwarf planet Pluto has a semi-major axis of about 5.9×10^{12} m and should not be affected by the term Λ. Since Λ has units of inverse length squared, we find the simple bound

$$\frac{1}{\sqrt{\Lambda}} \gg 5.9 \times 10^{12}\,\text{m} \quad \text{or} \quad \Lambda \ll 2.9 \times 10^{-26}\frac{1}{\text{m}^2}. \tag{5.108}$$

The observed value of Λ is about 10^{-52}m^{-2}, much smaller than the bound set by the Solar System. In other words, the contribution of Λ becomes only significant on cosmological scales.

Weak gravity

Exercise 2.14. We begin with $\bar{h}_{ab} = h_{ab} - \eta_{ab}h/2$ and take the trace of this equation which gives $\bar{h} = h - 4d/2 = -h$. Here we used $\eta^{ab}\eta_{ab} = \delta^a_a = 4$. Now we can substitute h by $-\bar{h}$ and arrive at $\bar{h}_{ab} = h_{ab} + \eta_{ab}\bar{h}/2$ which is equivalent to $h_{ab} = \bar{h}_{ab} - \eta_{ab}\bar{h}/2$.

Exercise 2.15. We start by recalling that $g_{ab} = \eta_{ab} + h_{ab}$ implies that $g^{ab} = \eta^{ab} - h^{ab}$, see Eq. (2.83). Next, instead of (2.91) we must consider

$$X'^i = X^i + \xi^i(X^k), \tag{5.109}$$

from which we directly compute the analogue of (2.93) which reads

$$\frac{\partial X'^i}{\partial X^j} = \delta^i_j + \frac{\partial \xi^i}{\partial X^j}. \tag{5.110}$$

Using, as in the main text, Eq. (1.34) gives

$$g'^{ab} = \eta^{ab} - h^{ab} + \partial^a \xi^b + \partial^b \xi^a. \tag{5.111}$$

Therefore,

$$h'^{ab} = h^{ab} - \partial^a \xi^b - \partial^b \xi^a, \tag{5.112}$$

which indeed matches Eq. (2.95) with lowered indices. Raising and lowering indices with η does not affect the signs. Computing the transformation property of the trace yields $h' = h - 2\partial_a \xi^a$ which we can combine with the previous equation to find the transformation of the trace-reverse tensor, we have

$$h'^{ab} - \frac{1}{2}\eta^{ab}h' = h^{ab} - \frac{1}{2}\eta^{ab}h - \partial^a \xi^b - \partial^b \xi^a + \eta^{ac}\partial_c \xi^b,$$

$$\Leftrightarrow \quad \bar{h}'^{ab} = \bar{h}^{ab} - \partial^a \xi^b - \partial^b \xi^a + \eta^{ac}\partial_c \xi^b. \tag{5.113}$$

Lastly, we can differentiate to arrive at

$$\partial_a \bar{h}'^{ab} = \partial_a \bar{h}^{ab} - \Box \xi^b, \tag{5.114}$$

where the last two terms cancelled because the order of partial differentiation does not matter.

Variational approach to General Relativity

Exercise 2.16. We begin with $f(x + \delta x) = f(x) + f'(x)\delta x + f''(x)(\delta x)/2 + \cdots$. We move $f(x)$ to the left-hand side and get

$f(x + \delta x) - f(x) = f'(x)\delta x + f''(x)(\delta x)/2 + \cdots$ and use the suggested notation so that

$$\delta f = f'(x)\delta x + f''(x)(\delta x)/2 + \cdots . \tag{5.115}$$

Now, we divide by δx and consider the limit $\delta x \to 0$ and we find that indeed $f'(x) = \lim_{\delta x \to 0} \delta f / \delta x$.

Exercise 2.17. To find the energy–momentum tensor we need to make the metric explicit in the action, we write $F^{ab}F_{ab} = F_{ab}F_{cd}g^{ca}g^{db}$ so that the action (without source terms) is given by

$$S = \int \frac{1}{16\pi} F_{ab}F_{cd}g^{ca}g^{db}\sqrt{-g}\,d^4x. \tag{5.116}$$

Therefore, the matter Lagrangian is $\mathcal{L}_{\mathrm{EM}} = F_{ab}F_{cd}g^{ca}g^{db}\sqrt{-g}$. Recall Eq. (2.118) for the square root term. We can now consider variations with respect to the metric

$$
\begin{aligned}
\delta\mathcal{L}_{\mathrm{EM}} &= \frac{F_{ab}F_{cd}}{16\pi}\left(\delta g^{ca}g^{db}\sqrt{-g} + g^{ca}\delta g^{db}\sqrt{-g} - \frac{1}{2}g^{ca}g^{db}\sqrt{-g}\,g_{mn}\delta g^{mn}\right) \\
&= \frac{F_{ab}F_{cd}}{16\pi}\left(\delta g^{ca}g^{db} + g^{ca}\delta g^{db} - \frac{1}{2}g^{ca}g^{db}g_{mn}\delta g^{mn}\right)\sqrt{-g} \\
&= \frac{1}{16\pi}\left(F_{nb}F_{md}g^{db} + F_{an}F_{cm}g^{ca} - \frac{1}{2}F_{ab}F_{cd}g^{ca}g^{db}g_{mn}\right)\sqrt{-g}\,\delta g^{mn} \\
&= \frac{1}{16\pi}\left(F_{nb}F_m{}^b + F_{an}F^a{}_m - \frac{1}{2}F_{ab}F^{ab}g_{mn}\right)\sqrt{-g}\,\delta g^{mn} \\
&= \frac{1}{16\pi}\left(2F_{nb}F_m{}^b - \frac{1}{2}F_{ab}F^{ab}g_{mn}\right)\sqrt{-g}\,\delta g^{mn} \\
&= -\frac{1}{8\pi}\left(F_{bn}F_m{}^b + \frac{1}{4}F_{ab}F^{ab}g_{mn}\right)\sqrt{-g}\,\delta g^{mn}. \tag{5.117}
\end{aligned}
$$

Finally, using Eq. (2.127) we arrive at

$$T_{mn} = -\frac{2}{\sqrt{-g}}\frac{\delta\mathcal{L}_{\mathrm{EM}}}{\delta g^{mn}} = \frac{1}{4\pi}\left(F_{bn}F_m{}^b + \frac{1}{4}F_{ab}F^{ab}g_{mn}\right), \qquad (5.118)$$

which is in agreement with Eq. (2.50).

Exercise 2.18. We know that the Ricci R already contains second partial derivatives of the metric. Naively, we would expect to arrive at fourth-order differential equations when computing the Euler–Lagrange equations because we need to integrate by parts twice when varying. However, the Einstein field equations are only second order because the variations δR_{ab} do not contribute to the field equations, see the discussion after Eq. (2.121).

The moment we consider a Lagrangian which is not linear in the Ricci scalar, this argument will not work. If our functional contains the term $R_{ij}R^{ij}$ then we would arrive at $\delta R_{ij}R^{ij} + R_{ij}\delta R^{ij}$. While δR_{ab} does give a surface term $\delta R_{ij}R^{ij}$ does not. The integration by parts used to arrive at Eq. (2.123) will now also act on the Ricci tensor R^{ij} introducing higher-order derivative terms. The same holds for gravitational actions containing an arbitrary function of the Ricci scalar which is not linear. All such theories yield fourth-order theories (or higher order if covariant derivatives are also included).

Additional discussion: This immediately leads to some problems when the Newtonian limit is considered. In Sec. 2.2.2, we discussed how the Ricci tensor is related to the Poisson equation of Newtonian gravity. Therefore, a higher order theory of gravity will not directly reduce to the Poisson equation but will contain higher derivative terms. These must be very carefully examined as any gravitational theory has to pass a series of stringent tests to be compatible with current observational data. It turns out that many gravitational theories already fail solar system tests, while those that do not tend to fail on cosmological scales. Constructing a gravitational theory as successful as General Relativity is indeed a very difficult task.

5.3.   Solutions: Schwarzschild Solutions

The Schwarzschild solution

Exercise 3.1. The Schwarzschild radius in physical units is $r_S = 2GM/c^2$. For the proton mass we take $m_{\mathrm{p}} = 1.67 \times 10^{-27}\,\mathrm{kg}$ so that

$$(r_s)_{\text{proton}} = \frac{2 \times 6.67 \times 10^{-11}\mathrm{m}^3\mathrm{kg}^{-1}\mathrm{s}^{-2} \times 1.67 \times 10^{-27}\mathrm{kg}}{(2.99 \times 10^8 \mathrm{ms}^{-1})^2}$$

$$= 2.49 \times 10^{-54}\,\mathrm{m}, \tag{5.119}$$

which is a very small number! The radius of the proton is about $r_{\mathrm{p}} = 0.84 - 0.87 \times 10^{-15}\,\mathrm{m}$ with some interesting discrepancies in recent measurements. This means that the Schwarzschild radius of the proton is almost 40 orders of magnitude smaller than its physical radius. This means we can safely neglect gravitational effects when studying particle physics.

Exercise 3.2. We work with $X^i = \{t, r, \theta, \phi\}$, $i = 0, 1, 2, 3$. We recall definition (1.71) of the Christoffel symbol. Let us begin with $n = 0$, then

$$\Gamma^0_{ij} = \frac{1}{2}g^{0k}\left(\partial_i g_{jk} + \partial_j g_{ki} - \partial_k g_{ij}\right)$$

$$= \frac{1}{2}g^{00}\left(\partial_i g_{j0} + \partial_j g_{0i} - \partial_0 g_{ij}\right) = \frac{1}{2}g^{00}\left(\partial_i g_{j0} + \partial_j g_{0i}\right), \tag{5.120}$$

where, in the first step, we took into account that the metric is diagonal, therefore g^{0k} is non-zero for $k = 0$ only. Then we used that the metric components are independent of time, hence $\partial_0 g_{ij} = 0$.

Therefore,

$$\Gamma^0_{ij} = -\frac{1}{2}e^{-A}\left(\partial_i g_{j0} + \partial_j g_{0i}\right). \tag{5.121}$$

The terms in the brackets can only give a non-zero contribution if either $j = 0, i = 1$ or $j = 1, i = 0$, thus

$$\Gamma^0_{01} = \Gamma^0_{10} = -\frac{1}{2}e^{-A}\partial_1 g_{00} = \frac{1}{2}e^{-A}A'e^A = \frac{1}{2}A'. \tag{5.122}$$

The other components are calculated along the same lines and the results are

$$\Gamma^1_{00} = \frac{1}{2}e^{A-B}A', \quad \Gamma^1_{11} = \frac{1}{2}B', \quad \Gamma^1_{22} = -e^{-B}r, \quad \Gamma^1_{33} = \sin^2\theta\Gamma^1_{22},$$

$$\Gamma^2_{12} = \Gamma^2_{21} = \frac{1}{r}, \quad \Gamma^2_{33} = -\sin(\theta)\cos(\theta), \tag{5.123}$$

$$\Gamma^3_{13} = \Gamma^3_{31} = \frac{1}{r}, \quad \Gamma^3_{23} = \Gamma^3_{32} = \cot(\theta).$$

Exercise 3.3. We begin with introducing standard spherical polar coordinates which gives

$$ds^2 = -\frac{\left(1-\frac{M}{2\rho}\right)^2}{\left(1+\frac{M}{2\rho}\right)^2}dt^2 + \left(1+\frac{M}{2\rho}\right)^4\left(d\rho^2 + \rho^2 d\theta^2 + \rho^2\sin^2\theta d\phi^2\right), \tag{5.124}$$

where we also changed the letter r to ρ in order to avoid confusion with the coordinate names.

Since our angular variables are already in the correct form, we need to match up to factors in front of the $d\Omega^2 = d\theta^2 + \sin^2\theta d\phi^2$. Therefore, we introduce a new radius as follows:

$$\left(1+\frac{M}{2\rho}\right)^4\rho^2 = r^2, \quad \text{or} \quad \left(1+\frac{M}{2\rho}\right)^2\rho = r. \tag{5.125}$$

This implies

$$dr = \left(1 + \frac{M}{2\rho}\right)^2 d\rho - \left(1 + \frac{M}{2\rho}\right)\left(\frac{M}{\rho}\right) d\rho \qquad (5.126)$$

$$= \left(1 + \frac{M}{2\rho}\right)^2 \left(\frac{2\rho - M}{2\rho + M}\right) d\rho. \qquad (5.127)$$

A direct calculation also shows that

$$1 - \frac{2M}{r} = \left(\frac{2\rho - M}{2\rho + M}\right)^2. \qquad (5.128)$$

Hence we find

$$\frac{dr^2}{\left(1 - \frac{2M}{r}\right)} = \left(\frac{2\rho + M}{2\rho - M}\right)^2 \left[\left(1 + \frac{M}{2\rho}\right)^2 \left(\frac{2\rho - M}{2\rho + M}\right) d\rho\right]^2 \qquad (5.129)$$

$$= \left(1 + \frac{M}{2\rho}\right)^4 d\rho^2, \qquad (5.130)$$

which shows that Eq. (3.66) really is equivalent to the Schwarzschild metric.

Exercise 3.4. In this question we will only state the results as some previous exercises were quite explicit about the computations required. Doing these calculations takes some time and patience. We use coordinates $X^i = \{t, r, \theta, \phi\}$ with $i = 0, 1, 2, 3$. The non-vanishing Christoffel symbol components are

$$\Gamma^0_{01} = \Gamma^0_{10} = A',$$

$$\Gamma^1_{00} = e^{2A-2B} A', \quad \Gamma^1_{11} = B', \quad \Gamma^1_{22} = -r - r^2 B', \quad \Gamma^1_{33} = \sin^2\theta \Gamma^1_{22},$$

$$\Gamma^2_{12} = \Gamma^2_{21} = \frac{1}{r} + B', \quad \Gamma^2_{33} = -\sin(\theta)\cos(\theta),$$

$$\Gamma^3_{13} = \Gamma^3_{31} = \frac{1}{r} + B', \quad \Gamma^3_{23} = \Gamma^3_{32} = \cot(\theta). \qquad (5.131)$$

These yield the following Ricci tensor components

$$R_{00} = e^{2A-2B} \left(\frac{2}{r} A' + (A')^2 + A'B' + A'' \right),$$

$$R_{11} = -\frac{2}{r} B' - (A')^2 + A'B' - A'' - 2B'',$$

$$R_{22} = -r^2 \left(\frac{1}{r} A' + \frac{3}{r} B' + A'B' + (B')^2 + B'' \right),$$

$$R_{33} = \sin^2\theta \, R_{22}.$$

(5.132)

Exercise 3.5. The Einstein tensor is defined by $G_{ab} = R_{ab} - g_{ab}R/2$. So, first we calculate the Ricci scalar

$$R = R^0{}_0 + R^1{}_1 + R^2{}_2 + R^3{}_3 = g^{00}R_{00} + g^{11}R_{11} + g^{22}R_{22} + g^{33}R_{33}.$$

(5.133)

Applying this to the Ricci tensor components from the previous exercise, we find

$$R^0_0 = -e^{-2B} \left(\frac{2}{r} A' + (A')^2 + A'B' + A'' \right),$$

$$R^1_1 = e^{-2B} \left(-\frac{2}{r} B' - (A')^2 + A'B' - A'' - 2B'' \right),$$

$$R^2_2 = -e^{-2B} \left(\frac{1}{r} A' + \frac{3}{r} B' + A'B' + (B')^2 + B'' \right),$$

$$R^3_3 = R^2_2.$$

(5.134)

Lastly, we add these four terms together and simplify as much as possible to arrive at

$$R = -2e^{-2B} \left(\frac{2}{r} A' + \frac{4}{r} B' + (A')^2 + (B')^2 + A'B' + A'' + 2B'' \right).$$

(5.135)

We will show explicitly how to compute G_{00} and only state the other results. Due to the factor of g_{00} in front of the Ricci scalar we note that R_{00} and $g_{00}R$ will have the same prefactor, hence

$$G_{00} = R_{00} - \frac{1}{2}g_{00}R$$

$$= e^{2A-2B}\left(\frac{2}{r}A' + (A')^2 + A'B' + A'' - \frac{2}{r}A' - \frac{4}{r}B'\right.$$

$$\left. - (A')^2 - (B')^2 - A'B' - A'' - 2B''\right)$$

$$= e^{2A-2B}\left(-\frac{4}{r}B' - (B')^2 - 2B''\right). \tag{5.136}$$

A particularly nice feature of G_{00} is that in this component the vacuum equation $G_{00} = 0$ will be independent of A, and therefore will only depend on the function B. This in turn implies that we can solve this differential equation and can immediately find B. Before doing so, we state the other components of the Einstein tensor. These are

$$G_{11} = 2A'\left(\frac{1}{r} + B'\right) + B'\left(\frac{2}{r} + B'\right),$$

$$G_{22} = r^2\left(\frac{1}{r}A' + \frac{1}{r}B' + (A')^2 + A'' + B''\right), \tag{5.137}$$

$$G_{33} = \sin^2\theta G_{22}.$$

Exercise 3.6. As already mentioned before, the vacuum equation $G_{00} = 0$ only depends on B' and B'', so we can introduce a new variable $b = B'$ to reduce the order of the differential equation by one which is then given by

$$2b' + \frac{4}{r}b + b^2 = 0. \tag{5.138}$$

Next, we can introduce a new function defined by $b = r/\beta(r)$, then (5.138) becomes

$$\beta' - \frac{3}{r}\beta = \frac{r}{2}, \tag{5.139}$$

where we multiplied by the factor $-\beta/2$. This is a general linear first-order ODE which we can solve using integrating factors. The solution if given by

$$\beta(r) = -\frac{r^2}{2} - C_1 r^3, \tag{5.140}$$

here C_1 is a constant of integration. Let us now go back to our original variable B. First,

$$b(r) = \frac{r}{\beta(r)} = -\frac{r}{C_1 r^3 + r^2/2} = -\frac{1}{C_1 r^2 + r/2}, \tag{5.141}$$

which we can integrate to find B so that

$$B(r) = 2 \log \left| C_2 \frac{1 + 2C_1 r}{r} \right|, \tag{5.142}$$

with C_2 being another constant of integration. The metric function is e^{2B}, and we arrive at

$$g_{11} = e^{2B} = \left(2C_1 C_2 + \frac{C_2}{r} \right)^4. \tag{5.143}$$

Next, we substitute our solution for $B(r)$ into the equation for G_{11} and solve for A', this gives

$$A' = \frac{4C_1}{4C_1^2 r^2 - 1}. \tag{5.144}$$

Integration yields

$$A(r) = \log \left| C_3 \frac{2C_1 r - 1}{2C_1 r + 1} \right|, \tag{5.145}$$

which gives us the other metric function to be

$$g_{00} = e^{2A(r)} = \left(C_3 \frac{2C_1 r - 1}{2C_1 r + 1} \right)^2. \tag{5.146}$$

The constant C_3 can always be set to one by rescaling the time coordinate. Next as r becomes very large the metric should approach Minkowski spacetime. This implies $2C_1 C_2 = 1$ or $C_2 = 1/(2C_1)$ and hence our metric is

$$ds^2 = -\left(\frac{2C_1 r - 1}{2C_1 r + 1}\right)^2 dt^2 + \left(1 + \frac{1}{2C_1 r}\right)^4 (d\rho^2 + \rho^2 d\Omega^2)$$

$$= -\left(\frac{1 - \frac{1}{2C_1 r}}{1 + \frac{1}{2C_1 r}}\right)^2 dt^2 + \left(1 + \frac{1}{2C_1 r}\right)^4 (d\rho^2 + \rho^2 d\Omega^2),$$

$$(5.147)$$

and we are left with one constant of integration which has to be connected to the mass parameter. We know from the weak field approximation that g_{11} has to be approximately $1 + 2M/r + \cdots$. This can only be achieved if we choose the $C_1 = 1/M$ which is indeed the choice consistent with the result given by (3.22).

Exercise 3.7. We begin by relabelling the time and radial coordinates by using τ and ρ instead of t and r, respectively. Then the metric becomes

$$ds^2 = -d\tau^2 + \left(\frac{\mu/3}{\rho - \tau}\right)^{2/3} d\rho^2 + \left(\frac{9\mu}{8}(\rho - \tau)^2\right)^{2/3} d\Omega^2. \quad (5.148)$$

The Schwarzschild metric contains the term $r^2 d\Omega^2$ which motivates the first coordinate transformation

$$r = ((9\mu/8)(\rho - \tau)^2)^{1/3}, \quad \text{or} \quad r^{3/2} = \frac{3}{2}\sqrt{\mu/2}(\rho - \tau), \quad (5.149)$$

which after differentiation yields

$$r^{1/2}dr = \sqrt{\mu/2}(d\rho - d\tau), \quad \Rightarrow \quad d\tau = d\rho - \sqrt{2/\mu}r^{1/2}dr. \quad (5.150)$$

Next, this needs to be substituted into Eq. (5.148) and we find

$$ds^2 = -\left(d\rho - \sqrt{2/\mu}r^{1/2}dr\right)^2 + \left(\frac{\mu}{3}\frac{3}{2}\sqrt{\mu/2}r^{-3/2}\right)^{2/3}d\rho^2 + r^2d\Omega^2$$

$$= -\left(d\rho^2 - 2\sqrt{2/\mu}r^{1/2}d\rho\,dr + (2/\mu)rdr^2\right) + \frac{\mu}{2r}d\rho^2 + r^2d\Omega^2$$

$$= -\left(1 - \frac{\mu}{2r}\right)d\rho^2 - \frac{2r}{\mu}dr^2 + 2\sqrt{\frac{2r}{\mu}}d\rho\,dr + r^2d\Omega^2. \qquad (5.151)$$

Let us now write the metric g_{ab} in matrix form to visualise the previous equation

$$g_{ab} = \begin{pmatrix} -\left(1 - \frac{\mu}{2r}\right) & \sqrt{\frac{2r}{\mu}} & 0 & 0 \\ \sqrt{\frac{2r}{\mu}} & -\frac{2r}{\mu} & 0 & 0 \\ 0 & 0 & r^2 & 0 \\ 0 & 0 & 0 & r^2\sin^2\theta \end{pmatrix}. \qquad (5.152)$$

It is now fairly clear that we aim to find coordinates which diagonalise this matrix. Since our angular coordinates and also the radial coordinate are already in the correct form, we must introduce a new time coordinate. Let us try the following transformation

$$\rho = t + f(r), \quad d\rho = dt + f'(r)dr,$$

$$\Rightarrow \quad d\rho^2 = dt^2 + 2f'dt\,dr + f'(r)^2dr^2, \qquad (5.153)$$

where the prime is the derivative of f with respect to r. Then our metric becomes

$$ds^2 = -\left(1 - \frac{\mu}{2r}\right)\left(dt^2 + 2f'dt\,dr + f'(r)^2dr^2\right) - \frac{2r}{\mu}dr^2$$

$$+ 2\sqrt{\frac{2r}{\mu}}\left(dt + f'(r)dr\right)dr + r^2d\Omega^2 \qquad (5.154)$$

$$= - \left(1 - \frac{\mu}{2r}\right) dt^2 + \left(2\sqrt{\frac{2r}{\mu}} f'(r) - \left(1 - \frac{\mu}{2r}\right) f'(r)^2 - \frac{2r}{\mu}\right) dr^2$$

$$+ 2\left(\sqrt{\frac{2r}{\mu}} - \left(1 - \frac{\mu}{2r}\right) f'(r)\right) dt\, dr + r^2 d\Omega^2. \tag{5.155}$$

In order to make this line element diagonal, we need to choose $f'(r)$ such that the $dt\, dr$ vanishes. This means

$$f'(r) = \sqrt{\frac{2r}{\mu}} \left(1 - \frac{\mu}{2r}\right)^{-1}. \tag{5.156}$$

One can integrate this directly, however, we do not need to know $f(r)$ explicitly as only $f'(r)$ enters the transformed metric. Lastly, we work out the transformed dr^2 term for which we get

$$\frac{2r}{\mu}\left(1 - \frac{\mu}{2r}\right)^{-1} - \frac{2r}{\mu} = \frac{2r}{\mu}\left(\left(1 - \frac{\mu}{2r}\right)^{-1} - 1\right)$$

$$= \frac{2r}{\mu}\left(\frac{\mu}{2r - \mu}\right) = \left(1 - \frac{\mu}{2r}\right)^{-1}. \tag{5.157}$$

Therefore, we arrive at the final result

$$ds^2 = - \left(1 - \frac{\mu}{2r}\right) dt^2 + \left(1 - \frac{\mu}{2r}\right)^{-1} dr^2 + r^2 d\Omega^2, \tag{5.158}$$

which is indeed the Schwarzschild metric, provided we set $\mu = 4M$.

Exercise 3.8. The field equations (3.10) and (3.11) with cosmological constant become

$$e^B + rB' - 1 - \Lambda r^2 e^B = 0, \tag{5.159}$$

$$-e^B + rA' + 1 + \Lambda r^2 e^B = 0. \tag{5.160}$$

Addition of both equations yields $A = -B$ as in the Schwarzschild case. Now, the first equation is equivalent to

$$e^B \frac{d}{dr}\left(r - re^{-B}\right) = \Lambda r^2 e^B, \quad \Rightarrow \quad r - re^{-B} = \frac{\Lambda}{3}r^3 + C, \tag{5.161}$$

where C is a constant of integration. We set $C = 2M$ and solve for e^{-B} which results in

$$e^{-B} = 1 - \frac{2M}{r} - \frac{\Lambda}{3}r^2. \tag{5.162}$$

This is the Schwarzschild–de Sitter or Kottler solution.

The Schwarzschild interior solution

Exercise 3.9. Without loss of generality we assume that $k = 1$. The energy density ρ_0 cannot be negative, one can always achieve that value by rescaling the radial coordinate. Let us introduce a new coordinate via $r = \sin \chi$ which gives

$$ds^2 = d\chi^2 + \sin^2\chi \left(d\theta^2 + \sin^2\theta d\phi^2\right). \tag{5.163}$$

On the other hand, let us start with the definition of the $\mathbb{S}^3$. We choose coordinates x_i, $i = 1, 2, 3, 4$, and have

$$x_1^2 + x_2^2 + x_3^2 + x_4^2 = 1. \tag{5.164}$$

We parameterise the 3-sphere using three angles as follows:

$$x_1 = \sin \chi \sin \theta \cos \phi, \tag{5.165}$$

$$x_2 = \sin \chi \sin \theta \sin \phi, \tag{5.166}$$

$$x_3 = \sin \chi \cos \theta, \tag{5.167}$$

$$x_4 = \cos \chi. \tag{5.168}$$

A direct but lengthy calculation yields

$$dx_1^2 + dx_2^2 + dx_3^2 + dx_4^2 = d\chi^2 + \sin^2\chi \left(d\theta^2 + \sin^2\theta d\phi^2\right). \tag{5.169}$$

Hence, we showed that the spatial part of the Schwarzschild interior solution has the geometry of a 3-sphere.

Exercise 3.10. The field equations with cosmological term are $G_{ab} + \Lambda g_{ab} = 8\pi T_{ab}$. This means we can readily state the $a = b = 0$ field equation by adding the term $\Lambda g_{00} = -\Lambda e^A$ to the left-hand side of Eq. (3.24). This gives

$$\frac{1}{r^2}e^{A-B}(e^B + rB' - 1) - \Lambda e^A = 8\pi\rho_0 e^A. \tag{5.170}$$

We move the Λ to the right-hand side and can jump to Eq. (3.28) which becomes

$$\frac{d}{dr}\left(r - re^{-B}\right) = (8\pi\rho_0 + \Lambda)r^2. \tag{5.171}$$

We can now integrate this equation and solve for e^{-B}. Using the standard mass definition $m(r) = 4\pi\rho_0 r^3/3$ we find

$$e^{-B} = 1 - \frac{2m(r)}{r} - \frac{\Lambda}{3}r^2 = 1 - 8\pi\rho_0/3r^2 - \frac{\Lambda}{3}r^2, \tag{5.172}$$

which we can also write as $1 - kr^2$ with $k = 8\pi\rho_0/3 + \Lambda/3$.

Exercise 3.11. For the Schwarzschild interior solutions $\rho(r) = \rho_0 = $ const. and $m(r) = (4\pi/3)\rho_0 r^3$ so that

$$M_p = 4\pi\rho_0 \int_0^R \frac{\bar{r}^2 d\bar{r}}{\sqrt{1 - (8\pi/3)\rho_0\bar{r}^2}}$$

$$= \frac{3R}{4}\left(\sqrt{\frac{R}{2M}}\arcsin\left(\frac{2M}{R}\right) - \sqrt{1 - \frac{2M}{R}}\right), \tag{5.173}$$

using standard integration techniques starting with $(8\pi/3)\rho_0 r^2 = \sin^2 u$.

Extra material: Interestingly, making a series expansion in M yields

$$M_p = M + \frac{3M^2}{5R} + \cdots. \tag{5.174}$$

Hence, in lowest-order approximation, both mass definitions agree. Moreover, we can interpret the quantity $M_p - M$ as the gravitational binding energy of the star.

Exercise 3.12. For the desired result to be true, we need to show that the integrand is $4\pi\rho_0 r^2$. This is a somewhat laborious task.

We recall Eqs. (3.32) and (3.39). We also need Eq. (3.45), however, we need to choose the constant differently to $\rho_0 + p_c$. This particular choice is motivated because gives $e^{A(0)} = 1$ at the origin. The exterior metric is the Schwarzschild which satisfies $e^A = e^{-B}$. We will choose our constant accordingly as our mass definition measures the mass at infinity. This gives

$$e^{A(R)} = \frac{C^2}{(\rho_0 + p(R))^2} = \frac{C^2}{\rho_0^2}, \tag{5.175}$$

$$e^{-B(R)} = \left(1 - \frac{8\pi}{3}\rho_0 R^2\right) = \left(1 - \frac{2M}{R}\right), \tag{5.176}$$

and leads to the choice $C^2 = \rho_0^2(1 - 2M/R)$.

Next, we will begin this calculation as follows:

$$\rho_0 + 3p(r) = \rho_0 + 3\rho_0 \frac{(3p_c + \rho_0)\sqrt{\bullet} - (p_c + \rho_0)}{3(p_c + \rho_0) - (3p_c + \rho_0)\sqrt{\bullet}}$$

$$= \rho_0 \frac{2(3p_c + \rho_0)\sqrt{\bullet}}{3(p_c + \rho_0) - (3p_c + \rho_0)\sqrt{\bullet}}, \tag{5.177}$$

$$\rho_0 + p(r) = \rho_0 + \rho_0 \frac{(3p_c + \rho_0)\sqrt{\bullet} - (p_c + \rho_0)}{3(p_c + \rho_0) - (3p_c + \rho_0)\sqrt{\bullet}}$$

$$= \rho_0 \frac{2(p_c + \rho_0)}{3(p_c + \rho_0) - (3p_c + \rho_0)\sqrt{\bullet}}. \tag{5.178}$$

We used the notation $\sqrt{\bullet}$ as shorthand for the entire square root expression $\sqrt{1 - (8\pi/3)\rho_0 r^2}$. At $r = R$ this becomes $\sqrt{\bullet} = \sqrt{1 - 2M/R}$. Therefore

$$\frac{\rho_0 + 3p(r)}{\rho_0 + p(r)} = \frac{(3p_c + \rho_0)\sqrt{\bullet}}{(p_c + \rho_0)}, \tag{5.179}$$

and we arrive at

$$(\rho_0 + 3p(r))e^{\frac{A}{2}}e^{\frac{B}{2}} = \frac{(3p_c + \rho_0)}{(p_c + \rho_0)}\sqrt{\rho_0}\sqrt{1 - \frac{2M}{R}}. \tag{5.180}$$

From Eq. (3.42) we can solve for $\sqrt{1 - 2M/R}$ which gives

$$\sqrt{1 - \frac{2M}{R}} = \frac{p_c + \rho_0}{3p_c + \rho_0}, \qquad (5.181)$$

so that we can finally write

$$(\rho_0 + 3p(r))e^{\frac{A}{2}}e^{\frac{B}{2}} = \rho_0, \qquad (5.182)$$

which proves the statement about the integrand. Therefore $M_\infty = M$.

Extra task: The motivated reader may wish to prove a more general statement, namely that $M = M_\infty$ is implied by the field equation. This means that this result is not specific to the Schwarzschild interior solution but applies to all such solutions, not just the constant density case, see Lightman *et al.* (1975, Problem 16.24).

Geodesics in Schwarzschild spacetime

Exercise 3.13. In Lightman *et al.* (1975), the authors suggest to work with outgoing Eddington–Finkelstein coordinates which is particularly suited to this problem. However, one can also attack this more directly working with the usual Schwarzschild coordinates. In Eq. (3.98), we found the redshift of the Schwarzschild spacetime. Assuming the observer to be far away from the Schwarzschild radius, we have

$$z = \frac{\lambda_{\text{obs}}}{\lambda_e} - 1 \approx \frac{1}{\sqrt{1 - \frac{2M}{r_e}}} - 1, \qquad (5.183)$$

where r_e (the radial location of the emitter) is the position of the radio commentator. We will now find $r_e(t)$ using the geodesic equations (3.51) and (3.54) with $L = -1$ (massive particle, the radio commentator is not massless, his signal is) and $\ell = 0$ because we consider radial motion only. Then

$$\frac{dr}{dt} = \frac{\dot{r}}{\dot{t}} = \frac{\sqrt{E^2 - \left(1 - \frac{2M}{r}\right)}}{E/\left(1 - \frac{2M}{r}\right)}$$

$$= \sqrt{1 - \frac{1}{E^2}\left(1 - \frac{2M}{r}\right)}\left(1 - \frac{2M}{r}\right) \approx \frac{r}{2M} - 1, \quad (5.184)$$

where we expanded near $r = 2M$. The term linear in r is independent of E. Separation of variables and integration yield

$$2M \log\left|\frac{r}{2M} - 1\right| = t + t_0, \quad (5.185)$$

where t_0 is a constant of integration. We now have

$$\exp\left(\frac{t + t_0}{2M}\right) = \frac{r}{2M} - 1, \quad (5.186)$$

from which we can find the approximate relationship $\exp((t + t_0)/2M) \approx 1 - 2M/r$. Substitution of this result into Eq. (5.183) gives

$$\frac{\lambda_{\text{obs}}}{\lambda_{\text{e}}} \approx \frac{1}{\sqrt{1 - \frac{2M}{r_{\text{e}}}}} \approx \exp\left(-\frac{t + t_0}{4M}\right). \quad (5.187)$$

Therefore, we conclude $\mu = 4M$.

Exercise 3.14. We are dealing with a massive particle on a radial geodesic, and so we can set $\ell = 0$ and $L = -1$. Then we are left with one geodesic equation, namely Eq. (3.54) which for this case becomes

$$\dot{r}^2 + \left(1 - \frac{2M}{r}\right) = E^2. \quad (5.188)$$

Our initial condition $\dot{r} = 0$ at $r = 10M$ fixes the constant E such that

$$E^2 = \left(1 - \frac{2M}{10M}\right) = \frac{4}{5}. \quad (5.189)$$

Consequently, we are left with

$$\frac{dr}{d\lambda} = -\sqrt{\frac{4}{5} - \left(1 - \frac{2M}{r}\right)} = -\frac{1}{\sqrt{5}}\sqrt{\frac{10M}{r} - 1}, \qquad (5.190)$$

where we chose the negative sign as the particle is moving towards the centre. The proper time λ_0 required to reach the centre is therefore given by

$$\lambda_0 = -\sqrt{5}\int_{r=10M}^{r=0}\frac{\sqrt{r}dr}{\sqrt{10M - r}}. \qquad (5.191)$$

In order to evaluate this integral, we use the substitution $r = 10M\sin^2 x$, $dr = 20M\sin x\cos x\, dx$, and get

$$\int\frac{\sqrt{r}dr}{\sqrt{10M - r}} = \int\frac{\left(\sqrt{10M}\sin x\right)20M\sin x\cos x\, dx}{\sqrt{10M - 10M\sin^2 x}}$$

$$= 20M\int\sin^2 x\, dx = 10M\int(1 - \cos(2x))dx$$

$$= 10M\left(x - \frac{1}{2}\sin(2x)\right). \qquad (5.192)$$

Here we neglected the constant of integration as we will evaluate a definite integral. Now, $r = 0$ corresponds to $x = 0$, while $r = 10M$ corresponds to $x = \pi/2$. Therefore,

$$\lambda_0 = 10\sqrt{5}M\left[x - \frac{1}{2}\sin(2x)\right]_{x=0}^{x=\pi/2} = 5\sqrt{5}M\pi. \qquad (5.193)$$

Exercise 3.15. For $\Lambda \leq 0$ the metric components are clearly regular, however, we still have to be careful since we are working with spherical coordinates which do not cover the entire sphere, we still need to be cautious at the poles. For $\Lambda > 0$ the metric is singular if $1 - (\Lambda/3)r^2 = 0$ which we use to define $r_\Lambda = \sqrt{3/\Lambda}$.

The most efficient way to find the geodesics equations is to start with the Lagrangian

$$L = -\left(1 - \frac{\Lambda}{3}r^2\right)\dot{t}^2 + \left(1 - \frac{\Lambda}{3}r^2\right)^{-1}\dot{r}^2 + r^2\dot{\theta}^2 + r^2\sin^2\theta\dot{\phi}^2. \quad (5.194)$$

If we now compare with Eq. (3.48), we notice that the only difference is the functional form of $f(r)$, otherwise we can use the same equations that followed. In particular, we have

$$E = f(r)\dot{t} = \left(1 - \frac{\Lambda}{3}r^2\right)\dot{t}, \quad (5.195)$$

$$\ell = r^2\dot{\phi}. \quad (5.196)$$

Hence, the final equation is derived from (3.53) with $\ell = 1$ and $L = -1$, this yields

$$\dot{r}^2 + \left(1 - \frac{\Lambda}{3}r^2\right), \quad (5.197)$$

which we can solve for $\dot{r}$ to get

$$\dot{r} = \pm\sqrt{E^2 - 1 + (\Lambda/3)r^2}. \quad (5.198)$$

We can fix the constant E by using the initial condition that the velocity at the origin is v, this gives

$$\dot{r} = \pm\sqrt{v^2 + (\Lambda/3)r^2}. \quad (5.199)$$

Separation of variables and integration yields

$$r_\Lambda \operatorname{arcsinh}\left(\frac{r}{vr_\Lambda}\right) = \lambda + \lambda_0,$$

$$\Rightarrow \quad r(\lambda) = vr_\Lambda \sinh\left(\frac{\lambda + \lambda_0}{r_\Lambda}\right), \quad (5.200)$$

where λ_0 is a constant of integration. We can set $\lambda_0 = 0$ which means that $r(0) = 0$. Therefore, $r(\lambda)$ is an increasing function which passes through all values of the radial variable and so reaches r_Λ when $\lambda = r_\Lambda \operatorname{arcsinh}(1/v)$. This is in analogy to the Schwarzschild radius, geodesics are well defined. However, when working with coordinate time we note that this would diverge as $r \to r_\Lambda$.

Testing General Relativity — the classical tests

Exercise 3.16. The standard equation of an ellipse is generally written in the form $x^2/a^2 + y^2/b^2 = 1$ where a and b are the semi-major and semi-minor axes which were chosen to coincide with the Cartesian axes. In order to show that Eq. (3.66) is indeed the equation of an ellipse, we need to consider the most general ellipse. Its centre does not have to coincide with the origin and moreover the major and minor axes do not have to be parallel to the Cartesian axes.

Recall that the sum of the distances between any point P on an ellipse to the two focal points F_1, F_2 is constant, and equal to the major axis. This means we can define an arbitrary ellipse by $|PF_1| + |PF_2| = 2a$. We will now choose the origin to coincide with the focal point F_1. Pythagoras theorem gives that the distance f between the centre of the ellipse and the focal points satisfies $f^2 + b^2 = a^2$. Let r be the position vector of the point P, then we can write our ellipse with one focal point at the origin as follows $|r| + |r - 2f\hat{f}| = 2a$, where f is the position vector of the other focal point. Now we introduce polar coordinates so that we write $r = (r\cos(\phi), r\sin(\phi))$ and $\hat{f} = (\cos(\varphi_0), \sin(\varphi_0))$. Here φ_0 is the angle which determines the direction of the second focal point. Since $|r| = r$ we arrive at

$$(2a - r)^2 = |r - 2f\hat{f}|^2$$

$$= (r\cos(\phi) - 2f\cos(\varphi_0))^2 + (r\sin(\phi) - 2f\sin(\varphi_0))^2$$

$$= r^2 + 4f^2 - 4rf\,(\cos(\phi)\cos(\varphi_0) + \sin(\phi)\sin(\varphi_0)).$$

$$(5.201)$$

Expanding out the left-hand side gives $4a^2 - 4ar + r^2$ so that the r^2 terms cancel, this yields

$$a^2 - ar = f^2 - rf\cos(\phi - \varphi_0),$$

$$\Rightarrow \quad \frac{1}{r} = -\frac{a}{f^2 - a^2} + \frac{f}{f^2 - a^2}\cos(\phi - \varphi_0),$$

$$\Leftrightarrow \quad \frac{1}{r} = \frac{a}{b^2} - \frac{\sqrt{a^2 - b^2}}{b^2}\cos(\phi - \varphi_0). \tag{5.202}$$

Let us state Eq. (3.66) explicitly again, this reads

$$\frac{1}{r} = \frac{M}{\ell^2} + C\cos(\phi + \phi_0). \tag{5.203}$$

Finally, we are able to identify the parameters accordingly. We must choose $\phi_0 = -\varphi_0 + \pi$ to get the correct sign, and

$$\frac{a}{b^2} = \frac{M}{\ell^2}, \tag{5.204}$$

$$\frac{\sqrt{a^2 - b^2}}{b^2} = C = \sqrt{\frac{E^2 - 1}{\ell^2} + \frac{M^2}{\ell^4}}. \tag{5.205}$$

Hence, we are indeed dealing with an ellipse.

Exercise 3.17. We will use a fairly rough estimate here. The radius of Jupiter ♃ is roughly one-tenth of the Sun's radius, while the mass of Jupiter is three orders of magnitude smaller

$$\Delta\phi_{\text{♃}} = \frac{4GM_{\text{♃}}}{c^2 R_{\text{♃}}} = \frac{4G(10^{-3}M_{\odot})}{c^2(10^{-1}R_{\odot})} = 10^{-2}\Delta\phi_{\odot}. \tag{5.206}$$

Therefore, light deflection near Jupiter is about two orders of magnitude weaker than light deflection near the Sun.

Exercise 3.18. It is well known that series can be added, multiplied, etc. So we should consider the different terms separately. Using the Binomial series $(1 + x)^\alpha = 1 + \alpha x + \cdots$ for the first term we get

$$\left(1 - \frac{2M}{r}\right)^{-1} = 1 + (-1)\left(-\frac{2M}{r}\right) + \cdots = 1 + \frac{2M}{r} + \cdots . \quad (5.207)$$

Next, we consider the term

$$\left(1 - \frac{2M}{r}\right)\left(1 - \frac{2M}{r_0}\right)^{-1} = \left(1 - \frac{2M}{r}\right)\left(1 + \frac{2M}{r_0} + \cdots\right)$$

$$= 1 + \left(\frac{2M}{r_0} - \frac{2M}{r}\right) + \cdots , \quad (5.208)$$

where we used the first approximation again. In order to evaluate the entire second term in the integral of Eq. (3.104) we write it as follows:

$$\left(1 - \frac{r_0^2}{r^2}\frac{1 - 2M/r}{1 - 2M/r_0}\right)^{-1/2} = \left(1 - \frac{r_0^2}{r^2}\left(1 + \left(\frac{2M}{r_0} - \frac{2M}{r}\right) + \cdots\right)\right)^{-1/2}$$

$$= \left(1 - \frac{r_0^2}{r^2} - 2M\frac{r_0^2}{r^2}\left(\frac{1}{r_0} - \frac{1}{r}\right)\right)^{-1/2} + \cdots$$

$$= \left(1 - \frac{r_0^2}{r^2}\right)^{-1/2}\left(1 + \frac{r_0}{r(r + r_0)}M + \cdots\right). \quad (5.209)$$

Therefore, the complete integrand gives

$$\left(1 - \frac{r_0^2}{r^2}\right)^{-\frac{1}{2}}\left(1 + \frac{2M}{r}\right)\left(1 + \frac{r_0}{r(r + r_0)}M + \cdots\right)$$

$$\doteq \left(1 - \frac{r_0^2}{r^2}\right)^{-\frac{1}{2}}\left(1 + \frac{2M}{r} + \frac{r_0}{r(r + r_0)}M + \cdots\right)$$

$$= \left(1 - \frac{r_0^2}{r^2}\right)^{-\frac{1}{2}}\left(1 + \frac{2M}{r} + \frac{r_0 M + rM - rM}{r(r + r_0)} + \cdots\right)$$

$$= \left(1 - \frac{r_0^2}{r^2}\right)^{-\frac{1}{2}}\left(1 + \frac{3M}{r} - \frac{M}{r + r_0} + \cdots\right). \quad (5.210)$$

This final equation is equivalent to Eq. (3.105).

The Schwarzschild radius

Exercise 3.19. From $u = t - f(r)$ we get $dt = du + f'dr$ and $dt^2 = du^2 + 2f'du\,dr + (f')^2 dr^2$. Then the de Sitter metric becomes

$$ds^2 = -\left(1 - \frac{\Lambda}{3}r^2\right)\left(du^2 + 2f'du\,dr + (f')^2 dr^2\right)$$

$$+ \left(1 - \frac{\Lambda}{3}r^2\right)^{-1} dr^2 + r^2 d\Omega^2$$

$$= -\left(1 - \frac{\Lambda}{3}r^2\right) du^2 - 2\left(1 - \frac{\Lambda}{3}r^2\right) f'du\,dr$$

$$+ \left(1 - \frac{\Lambda}{3}r^2\right)\left(\left(1 - \frac{\Lambda}{3}r^2\right)^{-2} - (f')^2\right) dr^2 + r^2 d\Omega^2.$$

$$(5.211)$$

We can now choose $f(r)$ so that the dr^2 term vanishes, then the $du\,dr$ term will reduce to a constant. We select

$$f'(r) = \frac{1}{1 - \frac{\Lambda}{3}r^2}, \qquad (5.212)$$

and our de Sitter metric in the new coordinates becomes

$$ds^2 = -\left(1 - \frac{\Lambda}{3}r^2\right) du^2 - 2du\,dr + r^2 d\Omega^2. \qquad (5.213)$$

Clearly, this metric is now regular at $r_\Lambda = \sqrt{3/\Lambda}$ as the metric and inverse metric are both regular near this radius. However, this metric is not regular everywhere. This is due to our use of spherical coordinates which cannot be used to describe the entire sphere, we face the usual problems near the poles.

To find the explicit form or $f(r)$, we need to integrate Eq. (5.212) which gives

$$f(r) = \sqrt{3/\Lambda}\, \text{arctanh}(r\sqrt{\Lambda/3}),$$

$$\text{or} \quad f(r) = r_\Lambda \, \text{arctanh}(r/r_\Lambda), \qquad (5.214)$$

where we neglected the constant of integration.

Exercise 3.20. The incoming Eddington–Finkelstein coordinate v is given by

$$v = t + r + 2M \log \left| \frac{r}{2M} - 1 \right|, \qquad (5.215)$$

so that

$$dt = dv - \frac{dr}{1 - 2M/r},$$

$$dt^2 = dv^2 - \frac{2dv\, dr}{1 - 2M/r} + \frac{dr^2}{(1 - 2M/r)^2},$$

$$\left(1 - \frac{2M}{r}\right) dt^2 = \left(1 - \frac{2M}{r}\right) dv^2 - 2dv\, dr + \frac{dr^2}{1 - 2M/r}. \qquad (5.216)$$

Then the Schwarzschild metric becomes

$$ds^2 = -\left(1 - \frac{2M}{r}\right) dv^2 + 2dv\, dr + r^2 d\Omega^2. \qquad (5.217)$$

Exercise 3.21. From Eq. (3.117) and the previous exercise we have

$$dv = dt + \frac{1}{1 - 2M/r} dr, \qquad (5.218)$$

$$du = dt - \frac{1}{1 - 2M/r} dr, \qquad (5.219)$$

so that by adding and subtracting these equations we find

$$dt = \frac{1}{2}(dv + du), \qquad (5.220)$$

$$dr = \frac{1}{2}\left(1 - \frac{2M}{r}\right)(dv - du). \qquad (5.221)$$

We square both equations and multiply by the factor in the Schwarzschild metric to get

$$\left(1 - \frac{2M}{r}\right) dt^2 = \frac{1}{4}\left(1 - \frac{2M}{r}\right)(dv^2 + du^2 + 2du\, dv), \qquad (5.222)$$

$$\left(1 - \frac{2M}{r}\right)^{-1} dr^2 = \frac{1}{4}\left(1 - \frac{2M}{r}\right)(dv^2 + du^2 - 2dudv). \quad (5.223)$$

Next we combine these and find

$$-\left(1 - \frac{2M}{r}\right) dt^2 + \left(1 - \frac{2M}{r}\right)^{-1} dr^2$$

$$= \frac{1}{4}\left(1 - \frac{2M}{r}\right)\left(dv^2 + du^2 - 2dudv - (dv^2 + du^2 + 2dudv)\right)$$

$$= -\left(1 - \frac{2M}{r}\right) dudv. \quad (5.224)$$

Finally, the Schwarzschild metric becomes

$$ds^2 = -\left(1 - \frac{2M}{r}\right) dudv + r^2 d\Omega^2, \quad (5.225)$$

where r is defined implicitly by the equation

$$v - u = 2r + 4M \log\left|\frac{r}{2M} - 1\right|. \quad (5.226)$$

Exercise 3.22. We begin with

$$U = -e^{-\frac{u}{4M}} = -\exp\left(-\frac{1}{4M}\left(t - r - 2M \log\left|\frac{r}{2M} - 1\right|\right)\right)$$

$$= -e^{-\frac{t}{4M}} e^{\frac{r}{4M}} \sqrt{\frac{r}{2M} - 1}, \quad (5.227)$$

and likewise

$$V = e^{\frac{v}{4M}} = \exp\left(\frac{1}{4M}\left(t + r + 2M \log\left|\frac{r}{2M} - 1\right|\right)\right)$$

$$= e^{\frac{t}{4M}} e^{\frac{r}{4M}} \sqrt{\frac{r}{2M} - 1}. \quad (5.228)$$

Therefore, we can compute

$$T = \frac{1}{2}(V + U) = \frac{1}{2}e^{\frac{r}{4M}}\sqrt{\frac{r}{2M} - 1}\left(e^{\frac{t}{4M}} - e^{-\frac{t}{4M}}\right)$$

$$= e^{\frac{r}{4M}}\sqrt{\frac{r}{2M} - 1}\,\sinh(t/4M), \tag{5.229}$$

$$X = \frac{1}{2}(V - U) = \frac{1}{2}e^{\frac{r}{4M}}\sqrt{\frac{r}{2M} - 1}\left(e^{t/4M} + e^{-\frac{t}{4M}}\right)$$

$$= e^{\frac{r}{4M}}\sqrt{\frac{r}{2M} - 1}\,\cosh(t/4M). \tag{5.230}$$

Strictly speaking this is only valid for $r > 2M$. If $r < 2M$ we need to be careful with the exponential of the logarithm. Also, for $r < 2M$ we notice from Eq. (3.19) that the roles of the radial and time coordinate change. Hence, for $r < 2M$ the coordinate transformation is given by

$$T = e^{\frac{r}{4M}}\sqrt{1 - \frac{r}{2M}}\,\cosh(t/4M), \tag{5.231}$$

$$X = e^{\frac{r}{4M}}\sqrt{1 - \frac{r}{2M}}\,\sinh(t/4M). \tag{5.232}$$

5.4. Solutions: Cosmology

Classical and Modern Cosmology

Exercise 4.1.

(i) Probably the simplest way out of this paradox is to assume that the universe is of finite age and that only finitely many stars can be observed in a given volume. If the number density of stars is sufficiently low, the night sky will be dark.

(ii) We could also consider an eternal universe by taking into account that stars themselves only have a finite lifetime. So, we would have a situation where new stars appear and old stars disappear. Provided the number density of the stars stays reasonably low, we would again have a dark sky at night.

(iii) Another possibility would be to have an eternal and infinite universe that is expanding. Due to the resulting redshift of the light, only a finite amount of light can reach an observer and therefore the night sky would again be dark.

Exercise 4.2. Let us consider an eternal universe. The second law of thermodynamics states that entropy cannot decrease $\dot{S} \geq 0$, which means that heat will flow from warmer to cooler objects. Therefore, a universe of infinite age should already have achieved thermal equilibrium. Clearly this is not the case as the universe contains various structure of differing temperature. Therefore, the universe cannot be infinitely old.

Exercise 4.3. As the calculation is straightforward and explicit examples were provided in earlier parts, only the results will be stated. Using $X^a = \{t, \rho, \theta, \phi\}$, $a = 0, 1, 2, 3$, the non-vanishing Christoffel symbol components are

$$\Gamma^0_{11} = \frac{\dot{a}a}{1 - k\rho^2}, \Gamma^0_{22} = \rho^2 \dot{a}a, \Gamma^0_{33} = \sin^2\theta \Gamma^0_{33},$$

$$\Gamma^1_{01} = \Gamma^1_{10} = \frac{\dot{a}}{a}, \Gamma^1_{11} = \frac{k\rho}{1 - k\rho^2}, \Gamma^1_{22} = -\rho(1 - k\rho^2), \Gamma^1_{33} = \sin^2\theta \Gamma^1_{22},$$

$$\Gamma^2_{02} = \Gamma^2_{20} = \frac{\dot{a}}{a}, \Gamma^2_{12} = \Gamma^2_{21} = \frac{1}{\rho}\Gamma^2_{33} = -\sin\theta\cos\theta,$$

$$\Gamma^3_{03} = \Gamma^3_{30} = \frac{\dot{a}}{a}, \Gamma^3_{13} = \Gamma^3_{31} = \frac{1}{\rho}, \Gamma^3_{23} = \Gamma^3_{32} = \cot\theta. \tag{5.233}$$

Exercise 4.4. Substitution of the Christoffel symbol components given by (5.233) into Eq. (1.164) and summing over the indices d and b yields

$$R_{00} = -3\frac{\ddot{a}}{a},$$

$$R_{11} = \frac{a^2(t)}{1 - k\rho^2} \left(\frac{\ddot{a}}{a} + 2\frac{\dot{a}^2}{a^2} + 2\frac{k}{a^2} \right),$$

$$R_{22} = \rho^2 a^2(t) \left(\frac{\ddot{a}}{a} + 2\frac{\dot{a}^2}{a^2} + 2\frac{k}{a^2} \right),$$

$$R_{33} = \sin^2\theta R_{22}. \tag{5.234}$$

Therefore, we can compute the Ricci scalar as follows:

$$R = R_0^0 + R_1^1 + R_2^2 + R_3^3$$

$$= 3\frac{\ddot{a}}{a} + 3 \left(\frac{\ddot{a}}{a} + 2\frac{\dot{a}^2}{a^2} + 2\frac{k}{a^2} \right) = 6\frac{\ddot{a}}{a} + 6\frac{\dot{a}^2}{a^2} + 6\frac{k}{a^2}, \tag{5.235}$$

which matches Eq. (4.16).

Exercise 4.5. The line element of $\mathbb{S}^2$ is $ds^2 = d\Omega^2 = d\theta^2 + \sin^2\theta d\phi^2$, so that $\det g_{ij} = \sin^2\theta$. Hence

$$V(\mathbb{S}^2) = \int_0^{2\pi} \left(\int_0^\pi \sin\theta d\theta \right) d\phi = \int_0^\pi \sin\theta d\theta \int_0^{2\pi} d\phi = 4\pi. \tag{5.236}$$

For the 3-sphere we found $ds^2 = d\chi^2 + \sin^2\chi d\Omega^2 = d\chi^2 + \sin^2\chi(d\theta^2 + \sin^2\theta d\phi^2)$, therefore $\det g_{ij} = \sin^4\chi \sin^2\theta$ and the volume will be given by

$$V(\mathbb{S}^3) = \int_0^\pi \sin^2\chi d\chi \int_0^\pi \sin\theta d\theta \int_0^{2\pi} d\phi = 2\pi^2. \tag{5.237}$$

We start to recognise the pattern for higher-dimensional spheres, let us denote the line element of $\mathbb{S}^n$ by $ds_{\mathbb{S}^n}^2$, then we find

$$ds_{\mathbb{S}^3}^2 = d\chi^2 + \sin^2\chi ds_{\mathbb{S}^2}^2 = d\chi^2 + \sin^2\chi d\Omega^2. \tag{5.238}$$

So, for the 4-sphere we just need another angle ξ, say, and we have

$$ds_{\mathbb{S}^4}^2 = d\xi^2 + \sin^2\xi ds_{\mathbb{S}^3}^2, \tag{5.239}$$

so that $\det g_{ij} = \sin^6\xi \sin^4\chi \sin^2\theta$. Hence

$$V(\mathbb{S}^4) = \int_0^\pi \sin^3\xi d\xi \int_0^\pi \sin^2\chi d\chi \int_0^\pi \sin\theta d\theta \int_0^{2\pi} d\phi = \frac{8\pi^2}{3}. \quad (5.240)$$

For the final part of the exercise, let us denote $I_n = \int_0^\pi \sin^n x dx$, then we rewrite the previous results as $V(\mathbb{S}^2) = 2\pi I_1$, $V(\mathbb{S}^3) = 2\pi I_2 I_1$, and $V(\mathbb{S}^4) = 2\pi I_3 I_2 I_1$. So for $\mathbb{S}^n$ we would have to evaluate

$$V(\mathbb{S}^n) = 2\pi I_{n-1} I_{n-2} \cdots I_1. \quad (5.241)$$

The integrals I_n are well known and there exist various formulae to find them explicitly. They are the reason for the appearance of the gamma function.

Exercise 4.6. For a matter-dominated universe $w = 0$, with $\Lambda = 0$ and $k = -1$, solving the field equations means solving Eq. (4.27) which in this case becomes

$$t - t_0 = \int \left[\frac{8\pi\rho_0}{3a} + 1\right]^{-1/2} da = \int \frac{\sqrt{a}\, da}{\sqrt{8\pi\rho_0/3 + a}}. \quad (5.242)$$

Instead of the trigonometric substitution used for the $k = 1$ case, we must use hyperbolic functions here. Start with

$$a = \frac{8\pi\rho_0}{3} \sinh^2(u/2),$$

$$da = \frac{8\pi\rho_0}{3} \sinh(u/2) \cosh(u/2)\, du, \quad (5.243)$$

so that our integral becomes

$$t - t_0 = \int \frac{\sqrt{8\pi\rho_0/3} \sinh(u/2)(8\pi\rho_0/3) \sinh(u/2) \cosh(u/2)\, du}{\sqrt{8\pi\rho_0/3}\sqrt{1 + \sinh^2(u/2)}}$$

$$= \frac{8\pi\rho_0}{3} \int \sinh^2(u/2)\, du = \frac{4\pi\rho_0}{3} \int (\cosh(u) - 1) du$$

$$= \frac{4\pi\rho_0}{3} (\sinh(u) - u). \quad (5.244)$$

As in the $k = 1$ case we cannot find $a(t)$ explicitly but can write the solution in parametric form

$$a = \frac{4\pi\rho_0}{3}(\cosh u - 1), \tag{5.245}$$

$$t = \frac{4\pi\rho_0}{3}(\sinh u - u). \tag{5.246}$$

This solution expands indefinitely which can be seen as follows. For large $u \gg 1$ we have $\cosh u = (e^u + e^{-u})/2 \approx e^u/2$ and $\sinh u = (e^u - e^{-u})/2 \approx e^u/2$, and also $\sinh u \gg u$ for $u \gg 1$. Then

$$a \approx \frac{4\pi\rho_0}{3}(e^u/2 - 1) \approx \frac{2\pi\rho_0}{3}e^u, \tag{5.247}$$

$$t \approx \frac{4\pi\rho_0}{3}(e^u/2 - u) \approx \frac{2\pi\rho_0}{3}e^u, \tag{5.248}$$

and therefore we can conclude that $a(t) \approx t$ for late times. We can see this behaviour in Fig. 4.1, $a(t)$ approaches the diagonal asymptotically.

Exercise 4.7. Equation (4.39) is independent of the pressure and therefore remains unchanged. On the other hand, Eq. (4.23) now becomes $\Lambda = 4\pi(\rho_E + 3p_E)$. The presence of pressure changes the value of the cosmological constant of the Einstein static universe.

Exercise 4.8. Throughout this calculation we assume quantities like δa to be small, and neglect terms higher than linear order. Also, we will include pressure and the equation of state $p = w\rho$ in order to also answer the next exercise as the approach is identical. We begin with

$$\frac{\dot{a}}{a} = \frac{\dot{\delta a}}{a_E - \delta a(t)} = \frac{\dot{\delta a}}{a_E} + \cdots, \tag{5.249}$$

$$\frac{\ddot{a}}{a} = \frac{\ddot{\delta a}}{a_E - \delta a(t)} = \frac{\ddot{\delta a}}{a_E} + \cdots. \tag{5.250}$$

Then the conservation equation (4.24) becomes

$$\dot{\delta\rho} = -3\frac{\dot{\delta a}}{a_E}(\rho_E + p_E) = -3\frac{\dot{\delta a}}{a_E}(1 + w)\rho_E, \quad (5.251)$$

which we can integrate to find that $\delta\rho \propto -\delta a(1+w)$. Equation (4.23) gives

$$\frac{\ddot{\delta a}}{a_E} = -\frac{4\pi}{3}(\delta\rho + 3\delta p) = -\frac{4\pi}{3}(1 + 3w)\delta\rho. \quad (5.252)$$

This implies

$$\ddot{\delta a} \propto (1 + 3w)(1 + w)\delta a, \quad (5.253)$$

which is a standard second-order equation with exponential solutions if $(1 + 3w)(1 + w) > 0$. For $w = 0$ we have $\ddot{\delta a} \propto \delta a$ which implies instability. We can find oscillating solutions if $(1 + 3w)(1 + w) < 0$ which implies the range $-1 < w < -1/3$. Unfortunately, the values of the equation of state parameter do not describe physical matter source. Matter of this type is often referred to as dark energy.

Exercise 4.9. See previous exercise.

Exercise 4.10. Inspection of Eq. (4.42) suggests the substitution $a = \sqrt{3k/\Lambda}\cosh x$ which yields

$$\sqrt{\frac{\Lambda}{3}}(t - t_0) = \int \frac{da}{\sqrt{a^2 - 3k/\Lambda}} = \int 1dx$$

$$= \operatorname{arccosh}\left(\sqrt{\Lambda/(3k)}a\right), \quad (5.254)$$

which we can solve for a. This immediately leads to the result.

Physical cosmology

Exercise 4.11. The function $E(z)$ for a spatially flat, matter-dominated universe without Λ is given by $E(z) = \sqrt{\Omega_{m0}}(1 + z)^{3/2}$. Since there are no other matter sources, we have $\Omega_{m0} = 1$ so that

$$t_{\text{universe}} = \frac{1}{H_0}\int_0^\infty \frac{dz}{(1 + z)(1 + z)^{3/2}}$$

$$= \frac{1}{H_0} \int_0^\infty (1+z)^{-5/2} dz = \frac{2}{3H_0}. \tag{5.255}$$

This looks similar to the distance of the particle horizon given by Eq. (4.104). This is not unexpected when comparing Eq. (4.103) with Eq. (4.78).

Exercise 4.12. As in the previous exercise, we have $E(z) = (1+z)^{3/2}$, hence

$$\chi = \frac{1}{a_0 H_0} \int_0^z \frac{d\bar{z}}{(1+\bar{z})^{3/2}} = \frac{2}{a_0 H_0} \left(1 - \frac{1}{\sqrt{1+z}} \right). \tag{5.256}$$

This leads to

$$d_L = (1+z)a_0\chi = \frac{2}{H_0} \left((1+z) - \sqrt{1+z} \right), \tag{5.257}$$

which is in agreement with our earlier result given by Eq. (4.80).

Exercise 4.13. Starting with the definition $\Omega = 8\pi\rho/(3H^2)$, we get

$$\frac{\Omega}{\Omega_0} = \frac{8\pi\rho/(3H^2)}{8\pi\rho_0/(3H_0^2)} = \frac{\rho}{\rho_0} \frac{H_0^2}{H^2}. \tag{5.258}$$

In the radiation-dominated epoch $\rho \propto 1/a^4$, $\rho/\rho_0 = a_0^4/a^4$. Since $\Lambda = 0$ we have $\Omega_\Lambda = 0$, and

$$k = a^2 H^2 (\Omega - 1) = a_0^2 H_0^2 (\Omega_0 - 1). \tag{5.259}$$

Putting all this together gives

$$\frac{(\Omega - 1)}{(\Omega_0 - 1)} = \frac{a_0^2 H_0^2}{a^2 H^2} = \frac{a_0^2}{a^2} \frac{\Omega}{\Omega_0} \frac{\rho_0}{\rho} = \frac{a_0^2}{a^2} \frac{\Omega}{\Omega_0} \frac{a^4}{a_0^4}$$

$$= \frac{a^2}{a_0^2} \frac{\Omega}{\Omega_0} = (1+z)^{-2} \frac{\Omega}{\Omega_0}. \tag{5.260}$$

Now we solve for Ω and find

$$\Omega = \frac{\Omega_0 (1+z)^2}{1 + z\Omega_0(2+z)}. \tag{5.261}$$

Therefore $A = 2$.

Next, from Eq. (5.258) we have

$$\frac{H^2}{H_0^2} = \frac{\rho}{\rho_0}\frac{\Omega_0}{\Omega} = \frac{a_0^4}{a^4}\frac{\Omega_0}{\Omega} = (1+z)^4\frac{\Omega_0}{\Omega}. \tag{5.262}$$

Then, using the previous result we have

$$\frac{H^2}{H_0^2} = (1+z)^4\frac{\Omega_0(1+z\Omega_0(2+z))}{\Omega_0(1+z)^2}$$

$$= (1+z)^2(1+z\Omega_0(2+z)), \tag{5.263}$$

which is the desired result.

Exercise 4.14. Recall the solution (4.38) for the $k = +1$ case. First, we will fix the constant t_0 by requiring that $a(0) = 0$, which implies $t_0^2 = 8\pi\rho_0/3$. The other time value where the scale factor vanishes is $t_{end} = 2\sqrt{8\pi/\rho_0/3}$. The angle α travelled for the geodesics is

$$\alpha = \int_0^{t_{end}} \frac{dt}{a(t)}. \tag{5.264}$$

Using a standard trigonometric substitution, we arrive at

$$\alpha = \arcsin\left(\sqrt{3/(8\pi\rho_0)}(t-t_0)\right)\Big|_{t=0}^{t=t_2}$$

$$= \arcsin(1) - (-\arcsin(1)) = 2\arcsin(1) = \pi, \tag{5.265}$$

which is exactly halfway around the universe.

Exercise 4.15. As in the previous exercise we have to evaluate

$$\alpha = \int_0^{t_{end}} \frac{dt}{a(t)}, \tag{5.266}$$

with $a(t)$ implicitly defined in Eqs. (4.32) and (4.33). Therefore, this is less straightforward than the radiation-dominated case. However, things will become very simple. Let us change the integration variable from t to the parameter u. We have $dt = (4\pi\rho_0/3)(1-\cos(u))du$ from

Eq. (4.33) while $a(u)$ is given by Eq. (4.32). This gives

$$\alpha = \int_{u(t=0)}^{u(t_{\text{end}})} \frac{(4\pi\rho_0/3)(1-\cos(u))du}{(4\pi\rho_0/3)(1-\cos(u))} = u\Big|_{u(t=0)}^{u(t_{\text{end}})}, \qquad (5.267)$$

and we are left to identify the parameter values correctly. As already discussed in the main text, $t = 0$ corresponds to $u = 0$ while t_{end} is attained when $u = 2\pi$. This implies that $\alpha = 2\pi$, which is exactly all the way around the universe.

Exercise 4.16. By definition we have

$$z = \frac{\lambda_o - \lambda_s}{\lambda_s} = \frac{6570 - 6563}{6563} = \frac{7}{6563} \approx \frac{7}{7000} = 10^{-3}. \qquad (5.268)$$

Therefore,

$$d_L = \frac{z}{H_0} = \frac{c \times 10^{-3}}{60\,\text{km/s}}\,\text{Mpc} = \frac{3 \times 10^8\,\text{m/s} \times 10^{-3}}{6 \times 10^4\,\text{m/s}}\,\text{Mpc}$$

$$= \frac{3}{6} \times 10\,\text{Mpc} = 5\,\text{Mpc}. \qquad (5.269)$$

With $d_L \approx d_A$ for $z = 10^{-3}$, and $d_A = R/\theta$ we get

$$R = \theta d_A = 4' \times 5\,\text{Mpc} = 4 \times \frac{2\pi}{60 \times 360} \times 5\,\text{Mpc}$$

$$\approx \frac{1}{10 \times 18}\,\text{Mpc} = \frac{100}{18}\,\text{kpc} \approx 5.5\,\text{kpc}. \qquad (5.270)$$

Inflation

Exercise 4.17. Start with $H = \dot{a}/a$. The cosmological field equations with scalar field source can be written as

$$3H^2 = 8\pi\left(\frac{1}{2}\dot{\phi}^2 + V(\phi)\right), \qquad (5.271)$$

$$2\dot{H} + 3H^2 = -8\pi\left(\frac{1}{2}\dot{\phi}^2 - V(\phi)\right). \qquad (5.272)$$

Their sum gives

$$2\dot{H} + 6H^2 = 16\pi V(\phi), \quad \Rightarrow \quad \dot{H} = 8\pi V(\phi) - 3H^2. \tag{5.273}$$

On the other hand, from (5.271) we can solve for $\dot{\phi}$ and get

$$\dot{\phi} = \sqrt{\frac{1}{4\pi}(3H^2 - 8\pi V(\phi))}. \tag{5.274}$$

Accordingly, we have $F(H, \phi) = 3H^2 - 8\pi V(\phi))$.

When $V = V_0$ we can separate variables in (5.273) and get

$$\frac{dH}{8\pi V_0 - 3H^2} = dt,$$

$$\Rightarrow \quad t - t_0 = \frac{1}{\sqrt{24\pi V_0}} \operatorname{arctanh}\left(H/\sqrt{8\pi V_0/3}\right), \tag{5.275}$$

where t_0 is the constant of integration. This gives

$$H(t) = \sqrt{8\pi V_0/3} \tanh\left(\sqrt{8\pi V_0 3}(t - t_0)\right). \tag{5.276}$$

To find $a(t)$ we recall $H = \dot{a}/a = \frac{d}{dt}\log(a)$. Now, we integrate $H(t)$ and find

$$\log(a(t)) = \frac{1}{3}\log\cosh\left(\sqrt{24\pi V_0}(t - t_0)\right) + C, \tag{5.277}$$

and so

$$a(t) = a_0\left(\cosh(\sqrt{24\pi V_0}(t - t_0))\right)^{1/3}, \tag{5.278}$$

where a_0 is another constant of integration. When $t \gg t_0$ we can write $\cosh(x) \approx 1/2\exp(x)$, therefore,

$$a(t) \approx a_0\left(\frac{1}{2}\exp(\sqrt{24\pi V_0}(t - t_0))\right)^{1/3}$$

$$= a_0\left(\frac{1}{2}\right)^{1/3}\exp\left[\sqrt{8\pi V_0/3}(t - t_0)\right]. \tag{5.279}$$

Thus, $\mu = \sqrt{8\pi V_0/3}$.

Exercise 4.18. The scalar field action is

$$S_{\text{matter}} = \int \left[-\frac{1}{2} g^{ab} \nabla_a \phi \nabla_b \phi - V(\phi) \right] \sqrt{-g}\, d^4x, \qquad (5.280)$$

and for now we are only interested in variations with respect to the metric. Recalling that $\delta\sqrt{-g} = -\sqrt{-g} g_{ab} \delta g^{ab}/2$ gives

$$\delta S_{\text{matter}} = \int \left[-\frac{1}{2} \delta g^{ab} \nabla_a \phi \nabla_b \phi \sqrt{-g} \right.$$
$$\left. - \left(-\frac{1}{2} g^{ab} \nabla_a \phi \nabla_b \phi - V(\phi) \right) \frac{1}{2} \sqrt{-g} g_{ab} \delta g^{ab} \right] d^4x.$$

We rearrange the terms and find

$$\delta S_{\text{matter}} = \int \frac{1}{2} \left[-\nabla_a \phi \nabla_b \phi + \frac{1}{2} \Box \phi g_{ab} + V(\phi) g_{ab} \right] \delta g^{ab} \sqrt{-g}\, d^4x. \qquad (5.281)$$

Therefore,

$$T_{ab} = -\frac{2}{\sqrt{-g}} \frac{\delta L_\phi}{\delta g^{ab}} \qquad (5.282)$$
$$= -\frac{2}{\sqrt{-g}} \frac{1}{2} \left[-\nabla_a \phi \nabla_b \phi + \frac{1}{2} \Box \phi g_{ab} + V(\phi) g_{ab} \right] \sqrt{-g}$$
$$= \nabla_a \phi \nabla_b \phi - g_{ab} \left(\frac{1}{2} \Box \phi + V(\phi) \right)$$
$$= \nabla_a \phi \nabla_b \phi + g_{ab} \mathcal{L}. \qquad (5.283)$$

Now, if $\phi = \phi(t)$ then $\nabla_a \phi = \partial_a \phi$ which is non-zero if and only if the derivative with respect to time is considered.

Exercise 4.19. This exercise contains in fact two parts. First, we need to find the Klein–Gordon equation for an arbitrary metric and the consider the Friedmann–Lemaître–Robertson–Walker metric. Our Euler–Lagrange equations are given by Eq. (1.229) with covariant derivatives. Hence

$$\frac{\partial \mathcal{L}}{\partial \phi} = -\frac{\partial V}{\partial \phi} = -V', \tag{5.284}$$

$$\frac{\partial \mathcal{L}}{\partial(\nabla_c \phi)} = -\frac{1}{2} g^{ab} \left(\delta_a^c \nabla_b \phi + \nabla_a \phi \delta_b^c \right) = -g^{ac} \nabla_a \phi, \quad (5.285)$$

$$\nabla_c \frac{\partial \mathcal{L}}{\partial(\nabla_c \phi)} = -g^{ac} \nabla_a \nabla_c \phi = -\nabla^a \nabla_a \phi = -\Box \phi. \tag{5.286}$$

Next, we must compute the explicit form of $\Box \phi$. We have

$$g^{ac} \nabla_a \nabla_c \phi = g^{ac} \nabla_a (\partial_c \phi) = g^{ac} (\partial_a \partial_c \phi - \Gamma_{ac}^d \partial_d \phi) = -\ddot{\phi} - 3H\dot{\phi}, \tag{5.287}$$

where we used the Christoffel symbol components (5.233). This gives final result $\ddot{\phi} + 3H\dot{\phi} + V' = 0$.

Exercise 4.20. Start with differentiation Eq. (4.127) with respect to time. This gives

$$3\dot{H}\dot{\phi} + 3H\ddot{\phi} \simeq -V''\dot{\phi}, \tag{5.288}$$

using the product rule on the left and the chain rule on the right. We solve for $3\ddot{\phi}$ and arrive at

$$3\ddot{\phi} \simeq -\frac{V''}{H}\dot{\phi} - 3\frac{\dot{H}}{H}\dot{\phi}. \tag{5.289}$$

We can solve Eq. (4.127) for $\dot{\phi}$ and substitute to get

$$3\ddot{\phi} \simeq \frac{V''}{3H^2}V' + \frac{\dot{H}}{H^2}V'. \tag{5.290}$$

Now, using Eq. (4.126), we see that the first term contains η and the second term contains $-\epsilon$. Now we find the desired equation

$$3\ddot{\phi} \simeq (\eta - \epsilon)V'. \tag{5.291}$$

The slow-roll approximation requires $\ddot{\phi} \ll 1$ and therefore, we note that $\eta \ll 1$ will guarantee that this is indeed satisfied.

Bibliography

Abbott, B. P. *et al.* (2016). Observation of gravitational waves from a binary black hole merger, *Phys. Rev. Lett.* **116**, 061102.

Ade, P. A. R. *et al.* (2015). Planck 2015 results. XIII. Cosmological parameters, arXiv:1502.01589.

Aldrovandi, R. and Pereira, J. G. (2013). *Teleparallel Gravity — An Introduction* (Springer, Heidelberg).

Ashtekar, A., Berger, B. K., Isenberg, J. and MacCallum, M. (2015). *General Relativity and Gravitation — A Centennial Perspective* (Cambridge University Press, Cambridge).

Bishop, R. L. and Goldberg, S. I. (1980). *Tensor Analysis on Manifolds* (Dover Publications, New York).

Blagojevic, M. (2001). *Gravitation and Gauge Symmetries* (CRC Press, Taylor & Francis, New York).

Blagojevic, M. and Hehl, F. W. (2012). *Gauge Theories of Gravitation — A Reader with Commentaries* (Imperial College Press, World Scientific, London).

Choquet-Bruhat, Y. (2008). *General Relativity and the Einstein Equations* (Oxford University Press, Oxford).

do Carmo, M. (1992). *Riemannian Geometry* (Birkhäuser, Boston).

Dodelson, S. (2003). *Modern Cosmology* (Academic Press, San Diego).

Eisenhart, L. P. (1997). *Riemannian Geometry* (Princeton University Press, Princeton).

Escher, M. C. (2015). Gallery — recognition & success, http://www.mcescher.com/gallery/recognition-success/. Last accessed on 30th June 2016.

Felice, A. D. and Tsujikawa, S. (2010). $f(R)$ Theories, *Living Rev. Relativ.* **13**, 3, doi:10.1007/lrr-2010-3, http://www.livingreviews.org/lrr-2010-3. Last accessed on 30th June 2016.

Frankel, T. (2012). *The Geometry of Physics* (Cambridge University Press, Cambridge).

Gorbunov, D. S. and Rubakov, V. A. (2011). *Introduction to the Theory of the Early Universe: Cosmological Perturbations and Inflationary Theory* (World Scientific, Singapore).

Hawking, S. W. and Ellis, G. F. R. (1973). *The Large Scale Structure of Space-Time* (Cambridge University Press, Cambridge).

Hogg, D. W. (1999). Distance measures in cosmology, eprint astro-ph/9905116.

Isham, C. J. (2001). *Modern Differential Geometry for Physicists* (World Scientific, Singapore).

Liddle, A. R. (2015). *An Introduction to Modern Cosmology* (John Wiley & Sons, Chichester).

Lightman, A. P., Press, W. H., Price, R. H. and Teukolsky, S. A. (1975). *Problem Book in Relativity and Gravitation* (Princeton University Press, Princeton), http://www.nrbook.com/relativity/. Last accessed on 30th June 2016.

Maggiore, M. (2007). *Gravitational Waves — Volume 1: Theory and Experiments* (Oxford University Press, Oxford).

Maluf, J. W. (2013). The teleparallel equivalent of general relativity, *Annalen Phys.* **525**, 339–357.

Misner, C. W., Thorne, K. S. and Wheeler, J. A. (1973). *Gravitation* (W. H. Freeman and Company, San Francisco).

Nakahara, M. (2003). *Geometry, Topology and Physics* (CRC Press, Taylor & Francis, New York).

NASA (1999). Mars climate orbiter, http://mars.jpl.nasa.gov/msp98/orbiter/. Last accessed on 30th June 2016.

Rowan-Robinson, M. (2004). *Cosmology* (Oxford University Press, Oxford).

Ryan, M. P. and Shepley, L. C. (1975). *Homogeneous Relativistic Cosmologies* (Princeton University Press, Princeton), http://wwwrel.ph.utexas.edu/Members/larry/RyanShepley.pdf. Last accessed on 30th June 2016.

Sotiriou, T. P. and Faraoni, V. (2010). $f(R)$ Theories of gravity, *Rev. Mod. Phys.* **82**, 451–497.

Stephani, H., Kramer, D., MacCallum, M. A. H., Hoenselaers, C. and Hertl, E. (2003). *Exact Solutions of Einstein's Field Equations* (Cambridge University Press, Cambridge).

Wald, R. M. (1984). *General Relativity* (The University of Chicago Press, Chicago).

Weinberg, S. (1972). *Gravitation and Cosmology: Principles and Applications of the General Theory of Relativity* (John Wiley & Sons, New York).

Weinberg, S. (2008). *Cosmology* (Oxford University Press, Oxford).

Will, C. M. (2014). The confrontation between general relativity and experiment, *Living Rev. Relativ.* **17**, 4, doi:10.1007/lrr-2014-4, http://www.livingreviews.org/lrr-2014-4. Last accessed on 30th June 2016.

Zwiebach, B. (2009). *A First Course in String Theory* (Cambridge University Press, Cambridge).

Index

Printed in the United States
By Bookmasters